Environmental Science, Engineering and Technology

Environmental Remediation Technologies, Regulations and Safety

nova
science publishers
www.novapublishers.com

Environmental Science, Engineering and Technology

Conserving, Halting and Reinstating Degraded Natural Resources during the UN Decade on Ecosystem Restoration
Rajani Srivastava, MSc, PhD and Pankaj Kumar, PhD (Editors)
2023. ISBN: 979-8-89113-067-8 (Hardcover)
2023. ISBN: 979-8-89113-140-8 (eBook)

Emerging Environmental Applications of Nanozymes
Seema Nara, PhD and Smriti Singh (Editors)
2023. ISBN: 979-8-88697-552-9 (Hardcover)
2023. ISBN: 979-8-88697-592-5 (eBook)

Modeling for Sustainable Development: A Multidisciplinary Approach
Rajendra Kumar, PhD, R. C. Singh, PhD, Rohit Khokher, PhD, and Vishal Jain, PhD (Editors)
2023. ISBN: 979-8-89113-015-9 (Softcover)
2023. ISBN: 979-8-89113-056-2 (eBook)

Pesticide Residues: Chemistry, Toxicology and Environmental Impact
David J. Goode (Editor)
2023. ISBN: 979-8-88697-557-4 (Softcover)
2023. ISBN: 979-8-88697-594-9 (eBook)

Unveiling Water's Implicit Journey in Domestic Interiors: A Closer Look at Different Approaches to Design
Ervin Garip, PhD, S. Banu Garip, PhD, Gözde Gökdemir, MA, and Uğur Efe Uçar, MA (Editors)
2023. ISBN: 979-8-88697-769-1 (Softcover)
2023. ISBN: 979-8-88697-826-1 (eBook)

More information about this series can be found at
https://novapublishers.com/product-category/series/environmental-science-engineering-and-technology/

Rajani Srivastava
and Pankaj Kumar
Editors

Conserving, Halting and Reinstating Degraded Natural Resources during the UN Decade on Ecosystem Restoration

nova
science publishers
www.novapublishers.com

Library of Congress Cataloging-in-Publication Data

ISBN: 979-8-89113-067-8

Published by Nova Science Publishers, Inc. † New York

To Our Parents

Contents

Preface

Ecosystem degradation is one of the ten threats officially recognized by the high-level board on threats, challenges, and change at the global level. The wealth of the habitats, their natural variety, and the commodities and services they can provide are all reduced as a result of this degradation, which has an impact on both native and migratory species. Many forms of environmental degradation exist. The ecology deteriorates when natural habitats are destroyed or when natural resources are used up. So there is a need for the reclamation of these biological systems. UN Environment Program (UNEP) and the Food and Agriculture Organization of the United Nations (FAO) are collaborating on the UN Decade on Ecosystem Restoration, which calls for the protection and recovery of the planet's natural ecosystems from one side to the other for the benefit of people and the environment. According to the UN, "the process of halting and reversing deterioration, resulting in increased ecosystem services and regained biodiversity" is what is meant by ecosystem restoration. Transformational ecosystem restoration demands a significant commitment from nations, the international community, civil society, and the private sector. The UN promotes the idea that several approaches to dealing with ecosystem restoration are necessary because what works beautifully in one place could fall short in another. Regenerative agriculture, ecological restoration, and rewilding are suggested methods.

In light of all these, this book will make a valuable contribution to the restoration of various ecosystems, including forests, grasslands, wetlands, wastelands, as well as some man-engineered ecosystems like agroecosystems. New and creative approaches to restoring all these ecosystems and assisting in the achievement of the 17 Sustainable Development Goals (SDGs) are also covered in various chapters. A chapter-by-chapter brief description is as follows:

In Chapter 1, Abaidya Nath Singh, Pankaj Kumar and others presented an overview of restoration and discussed the concept and strategies of restoration. Rajani Srivastava and Shreya Pandey in Chapter 2 discussed the causes and consequences of degradation. They emphasized the current problem of degradation by adopting such measures that not only contribute to the restoration process but are also helpful in the achievement of SDGs.

Chapter 3 written by Basant K. Tiwari, presents a computational analysis of complex ecological networks by using R. These computational analyses of ecological networks may provide useful insights into the inherent complexity of an ecosystem. Chapter 4 deals with forest ecosystem degradation and suggests a different sustainable approach for its restoration. This chapter is written by Sunil Singh and Rajani Srivastava and they also discussed many challenges (like high cost, shortage of regeneration materials, etc.) faced by ecologists during the process of restoration. Chapter 5 and 6 describes microbial tools and biotechnological tools for the restoration of degraded ecosystems. These two chapters are written by Neelam Atri and her co-workers and Richa Raghubanshi and associates.

Chapter 7 is written by Bhawana Raj, Rachna and Umesh Kumar Singh and describe the process of phytoremediation for the reclamation of pesticides and pharmaceutical-contaminated ecosystem. This chapter also discusses the flaws of existing conventional clean-up techniques and shifts towards ecological-based strategies. Chapter 8 highlights the ecological and economic importance of wetlands and its classification, and focuses on growing threats to wetlands by urbanization, agricultural practices, livestock grazing, and pollution. Here, methods of restoration are also discussed to reverse and protect these delicate ecosystems.

Chapter 9 covers the topic of climate-smart methods for sustainable agriculture. According to the authors of this chapter- climate-smart agriculture is a strategy to address the challenges of climate change and food security by sustainably increasing productivity, strengthening resilience, reducing GHG emissions, and increasing chances of achievement of national food security and sustainable development goals. This chapter was written by Nandita Ghoshal and her co-workers. Grasslands are crucial for raising cattle production and protecting the environment. Agricultural intensification, plantation forestry, urban growth, and mining, along with invasive non-native flora and rural sprawl, are the main factors contributing to the conversion and degradation of the grasslands. Chapter 10 deals with the degradation and restoration of grassland presented by M. Chalimalla and Rajani Srivastava.

Chapter 11 deals with wasteland and its reclamation through landfills, discussed by Vijai Krishan. Chapter 12 covers the wide area of computer simulation models (CSMs) in solving environmental problems. This chapter does not talk comprehensively or exhaustively about CSMs but also makes a way forward in the understanding of the models that are used these days in the scientific community.

Chapter 13 written by Saurabh Singh and J. P. Verma, discussed the method of reclamation of degraded lands through bioenergy plantations. They emphasized bioenergy plantations as an important tool for restoration and it is a critical factor that needs to be evaluated to ensure their long-term viability. Sustainability analysis involves the assessment of environmental, social, and economic impacts associated with bioenergy plantations. Bioenergy plantations can help restore degraded lands by different mechanisms depending upon the type of degradation of land. Chapter 14 describes the reclamation and restoration of soil through microbes. These microbes (like prokaryotes, fungi, bacteria, etc.) can clean up environmental contaminants and be helpful in the restorative process.

Dr. Rajani Srivastava
Dr. Pankaj Kumar

Acknowledgments

We bow down before the Almighty and our Parents for their grace which made things possible, we are also indebted to our teachers, and beloved family members who had faith and always stood with us in the realization of our dreams. First of all, our heartfelt thanks to our teacher Prof. K. P. Singh for his constant encouragement and support. We are most grateful to Director, Prof. A. S. Ragubanshi, IESD and Professor in Charge RGSC, BHU for giving us bright opportunities and moral support. We express our sincere thanks to Professor S. K. Tripathi, Aizawl Mizoram University and Professor Prashant Bhushan, HOD, Department of Ophthalmology, BHU for helping and encouraging us in writing this book. We also express our deepest gratitude to Prof. Nandita Ghoshal, Dr. P. C. Abhilash and Prof. Kavita Shah for their continuous encouragement at every step of academic activities.

We would also like to owe my sincere thanks to our friends Dr. Kalpana Awasthi, Assistant Professor, KN Govt. PG College, Bhadohi, and Dr. Sunita Singh, Assistant Professor, South Bihar, Gaya for their constant encouragement and support during all phases of book writing.

We express our sincere thanks to all authors Dr. A. N. Singh, Dr. Neelam Atri, Dr. Richa Raghubanshi, Dr. Umesh Singh, Dr. Basant Kumar Tiwari, Dr. Rajesh Rai, Dr. Sonam Gupta, Dr. Ajay Kumar Mishra, Dr. Alok, Dr. Ambrish, Dr. Ajay, Dr. Saurabh, Dr. Vijai Krishna, Dr. Sunil, Ms. Shreya Pandey and Mr. Mahesh, who worked hard for inculcating the new ideas in their chapters. Dr. Rajani wishes to express her deep sense of gratitude to her husband Mr. Pankaj Kumar Srivastava, her daughters Urvi and Kavya, and her sisters Ragini Srivastava Dr. Rekha Srivastava and Sadhana Srivastava for their constant affection, inspiration, and encouragement to put their best efforts for achieving betterment in life.

We also wish to extend our thanks to all those people, whose names we have not written, but whose contribution to this book was very much essential.

We sincerely thank all who helped us, directly and indirectly, to bring out this humble piece of work.

Dr. Rajani Srivastava
Dr. Pankaj Kumar
(Editors)

Chapter 1

Ecological Restoration: Concepts and Strategies

Abaidya Nath Singh[1]
Pankaj Kumar[1,*]
Arvind Pratap Singh[1]
and Sippy Dasani[2]

[1]Department of Botany, Udai Pratap College (Autonomous Institution), Varanasi, India
[2]Department of Botany, Bipin Bihari College, Jhansi, India

Abstract

There are many ways to use the word "restoration," but it often refers to going back to something's initial state. It should be viewed as encompassing entire ecosystems in ecological restoration. It is important to keep in mind that there are restoration alternatives that can be more realistic, such as replacement or rehabilitation. The habitat's physical and chemical characteristics, as well as the species themselves, are restoration components. Natural restorative mechanisms should be used wherever feasible, and after the degrading effects have been removed, they could even be adequate. However, each of them may require a specific treatment plan. Because restoration is an ongoing process, specifying success criteria can be challenging. The most key component is that ecosystem growth should go unhindered in the upward direction. From this, it follows that a successful restoration is a significant test of our knowledge of ecology.

Keywords: ecological restoration, rehabilitation, recuperation, reclamation

---

* Corresponding Author's Email: pankaj74bhu@gmail.com.

In: Conserving, Halting and Reinstating Degraded Natural Resources …
Editors: Rajani Srivastava and Pankaj Kumar
ISBN: 979-8-89113-067-8

1. Introduction

The primary causes of ecological degradation are disturbances caused by the changing climate and high-intensity human activities. In general, failing to recognize the true benefit of the beneficial functions provided by such systems is a major cause of ecosystem deterioration and loss (Crooks & Turner, 1999). For a very long time, humans have looked for ways to slow down or stop environmental degradation. Ecological restoration is the use of management strategies with the common aim of restoring the function of the pre-disturbed ecosystem, but the word has expanded to include a wider range of management interventions to restore ecosystem functioning, especially when the pre-disturbance conditions are unattainable (Hobbs et al., 2009; National Research Council, 1992). The strength of contemporary ecology is that it has been able to establish certain concepts that apply to all communities, although one community may differ from another due to climate, topography, or any of the many other factors.

Wilson famously penned 31 years ago: The next century will, I believe, be the era of restoration in ecology (Wilson, 1992). Ecological restoration is now viewed as a worldwide issue in research, politics, business, and practice—both professional and amateur (Aronson & Alexander, 2013a; Aronson & Alexander, 2013b). Ecological restoration seems to be one of the best efficient strategies for restoring the functionality and integrity of ecosystems in regions where degradation and change have gone too far or gone wrong (Balaguer et al., 2014; Devoto et al., 2012; Young et al., 2005). Furthermore, it seems to serve as a unifying idea and a hub for stakeholders and interest groups from many different backgrounds that are searching for fresh paradigms, approaches, and models (Murcia & Aronson, 2014). These entice us to discuss the concept and strategies of restoration ecology. Structure and function are the two main characteristics of an ecosystem, and each is composed of several components. They may be used to define and illustrate the damage that ecosystems can suffer (Bradshaw, 1987; Magnuson et al., 1980) (Figure 1). High values for both are frequently seen in an original ecosystem; however, this is not always the case. One or both traits are diminished, frequently to almost nothing, when something is degraded. Primary succession, a natural process, will return the ecosystem to its initial state if the region is let to develop naturally (Broady, 1993). The different kinds of natural ecosystems we observe today were first created by these processes, without any assistance (Bradshaw, 2002).

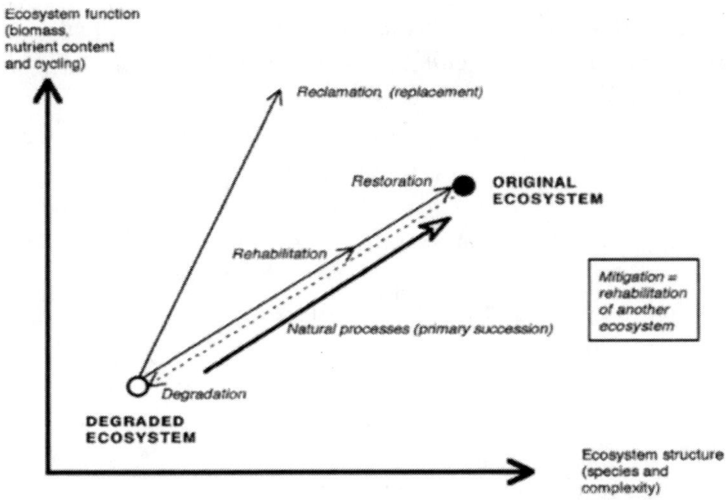

Figure 1. The two main features of structure and function may be used to represent the many choices for restoring a damaged ecosystem. Although not always equal but both attributes are typically diminished when deterioration takes place. Restoration, as used in this context, refers to returning the ecosystem to its original or former state regarding both structure and function. There are also a variety of other options, such as partial rehabilitation and replacement of the original item with something else. According to many, the term "reclamation" encompasses all these choices (Redrawn from Bradshaw, 1996).

2. Restoration-Meaning and Application

Two key elements—structure and function—can be used to summarize an ecosystem's characteristics. When these elements are combined to make two axes, the changes that are occurring in these may be graphically represented. Both require restoration because they have been damaged.

There is a chance to partially restore the functionalities, but it will be challenging to recreate the original structure. True restoration may therefore be both unrealistic and expensive. Reclamation and rehabilitation are therefore now the most popular choices. Both can deliver an endpoint that is more useful than the one already in place. The assumption is that a completely recovered ecosystem will be self-sustaining and resilient, that is, able to recover fully from changes and stresses that are to be expected.

The relevant definition of restoration is the act of restoring to a former state or position or an unimpaired or perfect condition. Bringing anything back to its original state or a healthy or vigorous state is what is meant by restoration. It entails both going back to an original state and going to a perfect, and healthy.

Restoring something to its prior state or condition is referred to as rehabilitation. Apart from the notion of perfection, this is equivalent to restoration. Something that has undergone rehabilitation is not anticipated to be in the original or healthy condition as something that has undergone restoration.

Making the land suitable for cultivation is known as reclamation. However, the definition of reclaim is to bring back to a proper state. There is no implication of returning to an original state but rather to a useful one.

3. Restoration Ecology Concept

An ecosystem that has been harmed, destroyed, or degraded can be helped to recover through a process known as ecological restoration (SERI, 2004; Suding et al., 2015). Ecological restoration tries to help recover a natural or seminatural ecosystem rather than imposing a new direction or shape, which is a key distinction between ecological restoration and other types of ecosystem repair. Consequently, an ecosystem is put on a trajectory of recovery through the activity of restoration so that it can survive and allow for the adaptation and evolution of its species (Gann et al., 2019; Higgs et al., 2018; McDonald et al., 2016).

Commonly, the term ecological restoration is used to refer to both a process (i.e., an action done to achieve a certain set of goals) and the goal of an ecosystem (i.e., its recovery). McDonald et al., (2016), who support the word "recovery," defined an ecological restoration effort as any action that aims to ultimately accomplish ecosystem recovery, insofar as practicable and relative to a suitable local native model (hence referred to as a reference ecosystem), regardless of the period necessary to reach the recovery objective.

3.1. Need for Principles and Standards

Ecosystem restoration is a challenging endeavor that takes much effort, money, and expertise. Despite having the best of intentions, many restoration projects and initiatives have fallen short of expectations (Gann et al., 2019). Ecological restoration significantly helps to conserve biodiversity and human well-being. The Standards acknowledge that better outcomes will result from suitable design, effective planning and execution of ample knowledge, skill, effort, and resources, a grasp of particular social contexts and concerns, appropriate stakeholder participation, and adequate monitoring for adaptive management. By defining requirements for technological implementation across various ecosystem types, the use of principles and standards can improve the efficacy of ecological restoration initiatives. Additionally, they offer a framework for both necessary and optional restoration that involves all relevant parties and considers sociocultural realities and requirements (Kim et al., 2021).

These principles can help agencies, institutions, or people who are involved in planning, implementing, and monitoring ecological restoration projects. They can also help regulators create agreements for mandated restoration and assess if those agreements have been reached (Clewell & Aronson, 2013). Finally, they can help policymakers when they are creating, approving, funding, and reviewing restoration projects of any kind. To decrease the risk of unintentional harm to ecosystems and native species, ecological restoration should be based on clear, thoughtful principles and standards. This will also make it easier to create high-quality projects and programs that can be monitored and evaluated.

4. Key Principles of Ecological Restoration

The restoration process is governed by three core concepts. The first is maintaining the integrity of the ecosystem. It is not recommended to attempt to develop an ecosystem that is distinct from the initial ecosystem type. For instance, one should not attempt to replace a dry deciduous forest with grassland or vice versa. The integrity of species is the second. It is important to ensure that only the proper species, those that are local or indigenous to the target environment, are introduced throughout the restoration process. The least possible intervention principle is the third. As much as possible, passive restoration procedures should be used, and only active approaches should be used in cases when passive restoration techniques are impractical. The emphasis should be on recovering ecological processes and achieving self-sufficiency in the ecosystem so that nature can manage the remaining aspects of restoration.

5. Components of Restoration

The five key components that need to be taken into account for ecological restoration are ecological, social, cultural, economic, and political (Aronson et al., 2010; Jackson et al., 1995). Understanding what it entails is crucial to choose what is worth trying. In essence, no matter what ecosystem is being tackled, after the destructive process has been stopped, the important element of any intervention is that it should relieve those conditions that are preventing the necessary reconstruction of the ecosystem. This might encompass a wide range of elements, depending on the ecosystem and the degree of deterioration.

Three things will need to be taken care of: (i) redesigning the habitat's physical features; (ii) reconstructing the chemical aspects, nutrients, and toxicity; and (iii) replacing absent species or eliminating undesired exotics.

What exactly is wrong will determine what is needed. To restore lost species, for instance, it may be necessary to reintroduce them. For example, acid land left over from coal mining will require liming; raw inert materials left to create soils will require nutrients; lakes overloaded with phosphorus may require mud pumping. It is crucial to recognize the influencing variables that will not receive better on their own and focus on them initially.

6. The Importance of the Reference Information

Ecological restoration involves copying nature by observing an interconnected natural system (Aber, 1987; Hough, 1984; MacMahon, 1987). The restoration aim is roughly modeled by a reference ecosystem (Kim et al., 2021; Lim et al., 2021). The reference model can be generated from a variety of sources, including information on the past, present, and conditions prevailing on or near the site, as well as information on predicted changes in environmental conditions that may result in changed biological assemblages, in the absence of adequate intact ecosystems of the same kind living nearby to the targeted location.

An ecological restoration project may be planned using a reference environment as a guide, and it can then be assessed using the reference ecosystem as a benchmark. The reference can be referred to as either the reference landscape or, if only a piece of a local landscape is to be restored, the reference landscape element in situations when the aim of restoration comprises two or more types of ecosystems.

The named ecosystem, landscape, or component may simply be referred to as the reference. The reference usually denotes a stage of development that is advanced and falls along the restoration's desired trajectory.

To put it another way, project objectives and strategies are created with the final intention that the restored ecosystem would mirror the characteristics of the reference. The reference might be a written description, one or more sites with model ecosystems, or a combination of both.

To put it another way, project objectives and strategies are created with the final intention that the restored ecosystem would mirror the characteristics of the reference. The reference might be a written description, one or more sites with model ecosystems, or a combination of both.

Both biotic and abiotic elements are included in the information collected about the reference (Gann et al., 2019; Higgs et al., 2018). Using the principles of an intact natural system, restoration ecology seeks to restore damaged ecosystems (Kim et al., 2021; Lim et al., 2021). Reference data is inherent in intact nature. In regions that have been maintained, such as national parks, we collected the reference data.

7. Attributes of Restored Ecosystems

When an ecosystem has enough biotic and abiotic resources to support continued growth without assistance or subsidies, it has recovered and is restored. It will be able to support itself structurally and functionally. It will show resilience against typical levels of environmental stress and disturbance. In terms of biotic and abiotic flow as well as cultural exchanges, it will engage with neighboring ecosystems (Gann et al., 2019; McDonald et al., 2016; SERI, 2004).

The following nine characteristics attributes are suggested by SERI as a framework for identifying when the restoration has been completed (Kim et al., 2012; SERI, 2004).

1. The restored ecosystem has a distinctive collection of species that coexist in the reference ecosystem and offer a suitable community structure.
2. To the maximum degree possible, the restored ecosystem is made up entirely of native species. In contrast, alien domesticated animals

and non-invasive ruderal and segetal species that likely coevolved with them can be tolerated in restored cultural environments. Segetals often coexist with crop species, whereas ruderals inhabit damaged areas.

3. All functional groups that are required for the continuous growth and/or stability of the restored ecosystem are present, or if they are not, the missing groups can naturally colonize the area.
4. The ecosystem after restoration has a physical habitat that can support populations of the species that are required for it to remain stable or develop along the planned course.
5. The repaired ecosystem does not show any evidence of malfunction and appears to function correctly for its ecological stage of development.
6. The restored ecosystem is properly integrated into a wider ecological matrix or landscape, with which it interacts through abiotic and biotic processes and exchanges.
7. Potential threats from the surrounding terrain to the restored ecosystem's health and integrity have been removed or minimized as much as practicable.
8. The ecosystem has been sufficiently repaired to be resilient enough to withstand the regular, recurring stress events in the immediate environment that preserve the ecosystem's integrity.
9. The restored ecosystem can survive eternally under the current environmental circumstances and is as self-sustaining as its reference ecology.

8. Strategies for Ecological Restoration

Here, six practical strategies are suggested to improve the efficiency and scale up ecological restoration efforts to accomplish the Decade's goals:

1. Be holistic, integrative, and comprehensive, especially when developing substantial plans for restorative action in mosaic landscapes. It is all about preserving native biodiversity as much as possible and protecting and regenerating natural capital, which includes native biodiversity and healthy ecosystems (Aronson et al., 2017; Aronson et al., 2020; Blignaut et al., 2014). An ecological restoration that takes a comprehensive approach aims to benefit

local communities, individuals who live near or in active areas, as well as people throughout the world (Clewell & Aronson, 2013). Working at a scale where social, cultural, and ecological issues interact is essential to a holistic approach.

2. When planning and carrying out ecological and associated restorative initiatives, consider and respect all legitimate forms of knowing, including Traditional Ecological Knowledge (Aronson et al., 2020).
3. Professionals concerned with ecological restoration and associated activities should collaborate closely with groups and organizations that are not necessarily involved in ecological restoration but work toward comparable objectives for ecological and human health, or when collaboration on seemingly unrelated objectives might result in synergistic or multiplier effects. Governments, companies, non-profit organizations, and academic institutions are a few examples of them (Aronson et al., 2020).
4. In the context of ecological restoration and linked restorative initiatives, develop and apply the scientific and technological understanding of soil microbiomes and biocrusts, as well as microbial communities and invertebrate assemblages in both marine and freshwater sediments. This strategy may be particularly useful in the massive stretches of arid, semi-arid, and desertified lands throughout the world that have already been severely or entirely degraded and are frequently heavily dependent on microbial communities for maintaining productivity through the supply of essential ecological processes such as nutrient cycling (Aronson et al., 2020; Blignaut et al., 2014; Moreira-Grez et al., 2019; Neilson et al., 2017; Soussi et al., 2016).
5. Provide early career professionals, practitioners, community leaders, academics, and administrators with much more on-site training and capacity-building opportunities at ecological restoration and restorative action sites (Aronson et al., 2020).
6. Highlight, research, and illustrate the interrelationships between restoring ecosystem health and trying to improve local and global human populations' mental, physical, social, and cultural health, as well as the general well-being and sustainability of communities, countries, and societies (Amberson et al., 2016; Aronson et al., 2016; Aronson et al., 2020; Elmqvist et al., 2015; Mills et al., 2017; Robinson et al., 2018; Speldewinde et al., 2015).

8.1. Planning for Restoration

Analyzing the degraded site is the first step in the restoration procedure. To determine whether restoration is necessary and whether any potential activities are needed, the site's existing circumstances should be carefully investigated. The causes that led to the degradation, the possibility that it may be stopped or mitigated, and the restoration methods should all be taken into consideration. During the initial planning phase, restoration should determine the essential resources (such as financing, human labor, technology, plant materials, and technical expertise) that have a direct impact on the project's viability. Early on in the planning phase and throughout implementation, support and participation from local governments and civic organizations may be quite helpful (Porensky et al., 2019).

Establishing attainable goals is an essential component of restoration planning. Although true goals are tailored to local ecological, social, legal, and economic contexts, pre-disturbance conditions, sometimes represented by adjacent reference sites, are generally used as restoration targets. Restoration initiatives are frequently driven by the desire to stop the future loss of protected populations or habitats. Controlling erosion, producing fodder on rangelands, preserving wildlife habitats, and maintaining cultural landscapes are some other objectives. It is now more crucial than ever to plan for climate change. After the project's objectives have been decided upon, the following stage is to lay down the project's goals and specify the success criteria. The direction and success of the project, as well as when and how restoration actions should be adjusted, may be determined by clearly defining the project's objectives and regularly evaluating quantifiable progress toward those objectives (Porensky et al., 2019).

8.2. Steps in Ecological Restoration

Before performing an ecological restoration, the steps listed in Figure 2 should be followed.

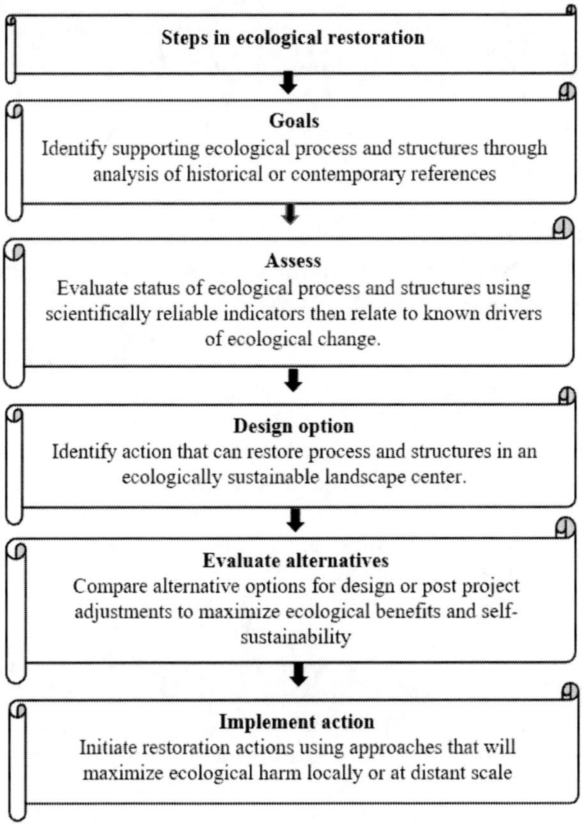

Figure 2. Ecological restoration Steps.

Conclusion

The restoration process can be seen as consisting of three phases: planning, implementation, and monitoring. The ecosystem restoration practices are developing, although at a very slow pace because the actors involved evaluate and modify their practices throughout the restoration process. In most cases, evaluation remains informal and is not documented. There is a need to share experiences and spread the word about finding both success and failure. In addition to traditional methods, digital participation should be encouraged. Another aspect is the economic analysis of the cost-efficiency of monitoring and evaluation. Increased knowledge in this respect has the potential to foster a better understanding of the significance of budgeting for evaluation in every restoration project. A fine balance between public perception and scientific rigor is highly warranted.

References

Aber, J. D. (1987). Restored forests and the identification of critical factors in species-site interactions. In W. R. Jordan, M. E. Gilpin & J. D. Aber (Eds.), *Restoration ecology A synthetic approach to ecological research* (pp. 241–250). Cambridge University Press.

Amberson, S., Biedenweg, K., James, J., & Christie, P. (2016). 'The Heartbeat of Our People': Identifying and measuring how salmon influences Quinault tribal well-being. *Society and Natural Resources*, *29*(12), 1389–1404. https://doi.org/10.1080/08941920.2016.1180727.

Aronson, J. (2010). What can and should be legalized in ecological restoration? *Revista Árvore*, *34*(3), 451–454. https://doi.org/10.1590/S0100-67622010000300009.

Aronson, J. C., Blatt, C. M., & Aronson, T. B. (2016). Restoring ecosystem health to improve human health and well-being: Physicians and restoration ecologists unite in a common cause. *Ecology and Society*, *21*(4), 39. https://doi.org/10.5751/ES-08974-210439.

Aronson, J., & Alexander, S. (2013a). Ecosystem restoration is now a global priority: Time to roll up our sleeves. *Restoration Ecology*, *21*(3), 293–296. https://doi.org/10.1111/rec.12011.

Aronson, J., & Alexander, S. (2013b). Steering towards Sustainability requires taking ecological restoration on board. *Natureza e Conservação [Nature and Conservation]*, *11*(2), 1–11. https://dx.doi.org/10.4322/natcon.2013.021.

Aronson, J., Blignaut, J. and Aronson, T. 2017. Conceptual frameworks and references for landscape-scale restoration: Reflecting back and looking forward. *Annals of the Missouri Botanical Garden.* 102(2):188-200, https://doi.org/10.3417/2017003.

Aronson, J., Goodwin, N., Orlando, L., Eisenberg, C., & Cross, A. T. (2020). A world of possibilities: Six restoration strategies to support the United Nation's Decade on Ecosystem Restoration. *Restoration Ecology*, *28*(4), 730–736. https://doi.org/10.1111/rec.13170.

Balaguer, L., Escudero, A., Martín-Duque, J. F., Mola, I., & Aronson, J. (2014). The historical reference in restoration ecology: Re-defining a cornerstone concept. *Biological Conservation*, *176*, 12–20. https://doi.org/10.1016/j.biocon.2014.05.007.

Blignaut, J. N., Aronson, J., & de Groot, R. (2014). Restoration of natural capital: A key strategy on the path to sustainability. *Ecological Engineering*, *65*, 54–61. https://doi.org/10.1016/j.ecoleng.2013.09.003.

Bradshaw, A. D. (1987). The reclamation of derelict land and the ecology of ecosystem. In: *Restoration Ecology: A Synthetic Approach to Ecological Research (eds.* W. R. Jordan, M. E. Gilpin & A. D. Aber) pp. 53–74. Cambridge University Press, Cambridge.

Bradshaw, A. D. 1996. Underlying principles of restoration. *Canadian Journal of Fisheries and Aquatic Sciences* 53(S1): 3–9. https://doi.org/10.1139/f95-265.

Bradshaw, A. D. (2002) Introduction and philosophy. In: *Handbook of Ecological Restoration*, vol. 1 (eds. M. R. Perrow and A. J. Davey), pp. 3–9. Cambridge University Press, Cambridge.

Broady, P. A. (1993). *Primary Succession on Land* Edited by J. Miles and D. W. H. Walton Special publication series of the British Ecological Society, number 12. Blackwell Scientific Publications, Oxford. 1993 309 pages. ISBN 0 632 03547. Antarctic Science, 5(3), 330-330. https://doi.org/10.1017/S0954102093250428.

Clewell, A. F., & Aronson, J. (2013). *Ecological restoration: Principles, values, and structure of an emerging profession* (2nd ed). Island Press. https://doi.org/10.5822/978-1-59726-323-8.

Crooks, S., & Turner, R. K. (1999). Integrated Coastal Management: Sustaining Estuarine Natural Resources. *Advances in Ecological Research*, *29*, 241–289. https://doi.org/10.1016/S0065-2504(08)60195-6.

Devoto, M., Bailey, S., Craze, P., & Memmott, J. (2012). Understanding and planning ecological restoration of plant–pollinator networks. *Ecology Letters*, *15*(4), 319–328. https://doi.org/10.1111/j.1461-0248.2012.01740.x.

Elmqvist, T., Setälä, H., Handel, S., van der Ploeg, S., Aronson, J., Blignaut, J., Gómez-Baggethun, E., Nowak, D., Kronenberg, J., & de Groot, R. (2015). Benefits of restoring ecosystem services in urban areas. *Current Opinion in Environmental Sustainability*, *14*, 101–108. https://doi.org/10.1016/j.cosust.2015.05.001.

Gann, G. D., McDonald, T., Walder, B., Aronson, J., Nelson, C. R., Jonson, J., Hallett, J. G., Eisenberg, C, Guariguata, M. R., Liu, J., Hua, F, Echeverría, C., Gonzales, E., Shaw, N., Decleer, K. & Dixon, K. W. (2019). International principles and standards for the practice of ecological restoration. *Restoration Ecology*, *27*(S1), S1-S46. https://doi.org/10.1111/rec.13035.

Higgs, E., Harris, J., Murphy, S., Bowers, K., Hobbs, R., Jenkins, W., Kidwell, J., Lopoukhine, N., Sollereder, B., Suding, K., Thompson, A., & Whisenant, S. (2018). On principles and standards in ecological restoration. *Restoration Ecology*, *26*(3), 399–403. https://doi.org/10.1111/rec.12691.

Hobbs, R. J., Higgs, E., & Harris, J. A. (2009). Novel ecosystems: Implications for conservation and restoration. *Trends in Ecology and Evolution*, *24*(11), 599–605. https://doi.org/10.1016/j.tree.2009.05.012

Hough, M. (1984). *City form and natural process*, *279*. Croom-Helm.

Jackson, L. L., Lopoukhine, N., & Hillyard, D. (1995). Ecological restoration: A definition and comments. *Restoration Ecology*, *3*(2), 71–75. https://doi.org/10.1111/j.1526-100X.1995.tb00079.x.

Kim, A. R., Lim, B. S., Seol, J., & Lee, C. S. (2021). Principle of restoration ecology reflected in the process creating the National Institute of Ecology. *Journal of Ecology and Environment*, *45*(1), 1–12. https://doi.org/10.1186/s41610-021-00187-w.

Kim, A. R., Lim, B. S., Seol, J., Lim, C. H., You, Y. H., Lee, W. S., & Lee, C. S. (2021). Diagnostic assessment and restoration plan for damaged forest around the Seokpo

Zinc Smelter, Central Eastern Korea. *Forests*, *12*(6), 1–26. https://doi.org/10.3390/f12060663.

Lim, C. H., Pi, J. H., Kim, A. R., Cho, H. J., Lee, K. S., You, Y. H., Lee, K. H., Kim, K. D., Moon, J. S., & Lee, C. S. (2021). Diagnostic evaluation and preparation of the reference information for river restoration in South Korea. *International Journal of Environmental Research and Public Health*, *18*(4), 1724. https://doi.org/10.3390/ijerph18041724.

MacMahon, J. A. 1987. Disturbed lands and ecological theory: an essay about a mutualistic association. In: Jordan III, W. R., M. E. Gilpin and J. D. Aber. 1987 (eds.). *Restoration Ecology a Synthetic Approach to Ecological Research*, pp. 221–237. Cambridge University Press, Cambridge.

Magnuson, J. J., H. A. Regier, W. J. Christie, and W. C. Sonzogni. 1980. *To Reinhabitat and Restore Great Lakes Ecosystems. The Recovery Process in Damaged Ecosystems*. Ann Arbor Science Publishers, Ann Arbor, Mich.

McDonald, T., Gann, G., Jonson, J., & Dixon, K. (2016). *International standards for the practice of ecological restoration – Including principles and key concepts* (1st ed) (pp. 11–24). Society for Ecological Restoration.

Mills, J. G., Weinstein, P., Gellie, N. J. C., Weyrich, L. S., Lowe, A. J., & Breed, M. F. (2017). Urban habitat restoration provides a human health benefit through microbiome rewilding: The microbiome Rewilding Hypothesis. *Restoration Ecology*, *25*(6), 866–872. https://doi.org/10.1111/rec.12610.

Moreira-Grez, B., Tam, K., Cross, A. T., Yong, J. W. H., Kumaresan, D., Nevill, P., Farrell, M., & Whiteley, A. S. (2019). The Bacterial Microbiome Associated with Arid Biocrusts and the Biogeochemical Influence of Biocrusts Upon the Underlying Soil. *Frontiers in Microbiology*, 10, 2143. https://www.frontiersin.org/articles/10.3389/fmicb.2019.02143/full.

Murcia, C., & Aronson, J. (2014). Intelligent tinkering in ecological restoration. *Restoration Ecology*, *22*(3), 279–283. https://doi.org/10.1111/rec.12100.

National Research Council. (1992). *Restoration of aquatic ecosystems: Science, technology and public policy*. National Academy Press.

Neilson, J. W., Califf, K., Cardona, C., Copeland, A., Van Treuren, W., Josephson, K. L., Knight, R., Gilbert, J. A., Quade, J., Caporaso, J. G., & Maier, R. M. (2017). Significant impacts of increasing aridity on the arid soil microbiome. *mSystems*, *2*(3), 1–15. https://doi.org/10.1128/mSystems.00195-16.

Porensky, L., Veblen, K. E., & Young, T. (2019). *Ecological Restoration. Encyclopedia Britannica*.

Robinson, J. M., Mills, J. G., & Breed, M. F. (2018). Walking ecosystems in microbiome-inspired green infrastructure: An ecological perspective on enhancing personal and planetary health. *Challenges*, *9*(2), 1–15. https://doi.org/10.3390/challe9020040.

Soussi, A., Ferjani, R., Marasco, R., Guesmi, A., Cherif, H., Rolli, E., Mapelli, F., Ouzari, H. I., Daffonchio, D., & Cherif, A. (2016). Plant-associated microbiomes in arid lands: Diversity, ecology and biotechnological potential. *Plant and Soil*, *405*(1–2), 357–370. https://doi.org/10.1007/s11104-015-2650-y.

Speldewinde, P. C., Slaney, D., & Weinstein, P. (2015). Is restoring an ecosystem good for your health? *Science of the Total Environment*, *502*, 276–279. https://doi.org/10.1016/j.scitotenv.2014.09.028.

Suding, K., Higgs, E., Palmer, M., Callicott, J. B., Anderson, C. B., Baker, M., Gutrich, J. J., Hondula, K. L., LaFevor, M. C., Larson, B. M., Randall, A., Ruhl, J. B., & Schwartz, K. Z. (2015). Conservation. Committing to ecological: Efforts around the globe need legal and policy clarification. *Science*, *348*(6235), 638–640. https://www.science.org/doi/10.1126/science.aaa4216.

Wilson, E. O. (1992). *The diversity of life*, *340*. Belknap Press.

Young, T. P., Petersen, D. A., & Clary, J. J. (2005). The ecology of restoration: Historical links, emerging issues and unexplored realms. *Ecology Letters*, *8*(6), 662–673. https://doi.org/10.1111/j.1461-0248.2005.00764.x.

Chapter 2

Ecosystem Degradation: Causes and Consequences

Rajani Srivastava*
and Shreya Pandey
Environmental Science (Environmental Technology),
Institute of Environment and Sustainable Development, RGSC,
Banaras Hindu University, Mirzapur, India

Abstract

Ecosystem degradation is a critical challenge faced by humankind. A major cause of degradation is anthropogenic activities. Degradation also led to the loss of ecosystem services. It is caused due to a variety of issues including pollution, overpopulation, deforestation, and global warming. Economic expansion and mindless consumption are the main causes that harm nature. Rapid industrialization and unplanned urbanization often advance in a way in developing countries; they have destructive effects on the environment, particularly on basic ecosystems, wildlife habitat and global biodiversity. Degradation harms human health, leading to ozone layer depletion, desertification and loss of biodiversity. Changing climate, land-use changes and anthropogenic activity all appear to be substantial contributors to biodiversity loss. To address the relevant problem, we must use resources wisely, promote sustainable development, encourage green concepts and most importantly, involve the community in all developmental endeavors. This chapter revolves around the cause and consequences of ecosystem degradation. To deal with the current scenario we need to adopt such

* Corresponding Author's Email: rajani.env@bhu.ac.in.

In: Conserving, Halting and Reinstating Degraded Natural Resources …
Editors: Rajani Srivastava and Pankaj Kumar
ISBN: 979-8-89113-067-8

measures that contribute significantly to the restoration of the ecosystem and Sustainable Development Goals.

Keywords: ecosystem degradation, global warming, desertification, sustainable development goals

1. Introduction

Ecosystem degradation is the deterioration of an ecosystem's structure, function, or decline in ecosystem services (Delgado & Marín, 2020). Economic progress depends on a robust and stable ecosystem. Many of the services like provisioning, regulating, cultural and supporting are derived by human beings directly or indirectly from natural ecosystems (Millennium Ecosystem Assessment, 2005). Local conflict, emigration and loss of a source of income can all be attributed to ecological degradation (Cao et al., 2020). The High-Level Threat Panel of the United Nations has listed ecosystem deterioration as one of its top 10 official warnings. Ecosystem degradation, which includes challenges like pollution, biodiversity loss, animal extinction, deforestation and desertification, and global climate change, is a highly significant issue on a global basis (Brown et al., 1987). For instance, the rapid consumption of limited natural resources brought on by industrialization and economic growth is the key factor in degradation (Chakravarty & Mandal, 2020). More profoundly and exhaustively than at any other point in human history, human activities have altered the ecosystem in recent decades (Pan et al., 2021). The global environment is significantly impacted by both rapid economic growth and population growth. Most of the population roughly three-fourths of the total population depends indirectly on natural resource-based activities for their livelihood, while the remaining population directly uses these resources for food, fuel, industrial output, and recreation (Maurya et al., 2020). Carbon dioxide emissions are widely used as an indicator of environmental degradation (Burki & Najam, 2021). As per O'Neill et al., (2010) in their analysis of 35 nations, slowed population growth might reduce carbon emissions by 1.4 to 2.5 billion tonnes annually by 2050. This would undoubtedly assist to address the climate issue. Land development and its associated activities have altered the structure and function of an ecosystem by bringing changes in ecosystem services (Lu et al., 2022). Rapid urbanization and agricultural growth are two examples of high-intensity land development that could result in the transformation of natural ecosystems into construction and farming land, impeding the ecological processes of the water cycle, soil formation and plant photosynthesis (Sun et al., 2021). The decline in ecosystem services tends to decrease the resilience capacity of linked socio-ecosystem (Gordon & Enfors, 2008). For economic growth and human flourishment in the world, natural resources must be managed sustainably.

The present chapter highlights various aspects that led to ecosystem degradation. An attempt has been made to emphasize on overall causes and consequences of ecosystem degradation.

2. Cause of Ecosystem Degradation

Ecosystems sustain a precise equilibrium. Multiple human activities posed a threat to upset this equilibrium and wipe out the planetary environment. Ecosystem degradation is just like a broad concept. It includes various issues as shown in Figure 1. The frequency and severity of natural hazards can change as a result of ecosystem degradation.

Some of the cause of ecosystem degradation includes:

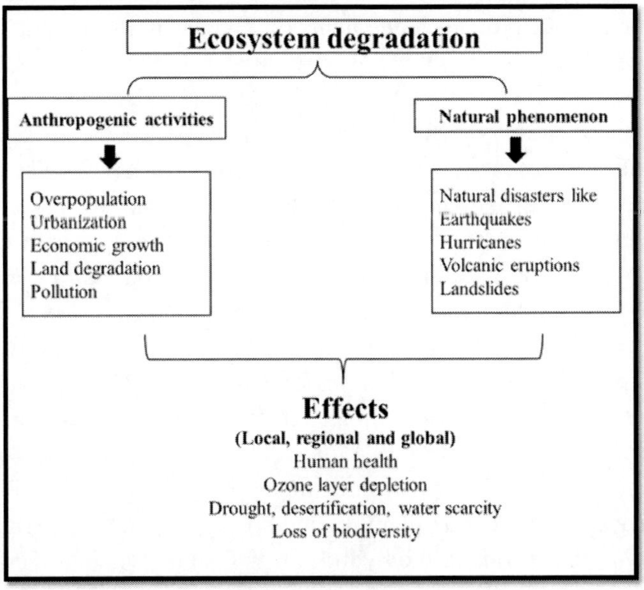

Figure 1. Different causes and effects of ecosystem degradation.

2.1. Land Degradation

One of the most alarming environmental issues that are faced by the planet is land degradation, which will only get worse if appropriate remedial action is not taken. Class I land, which makes up just around 3% of the world's land surface, is not found in the tropics. 8% more of the land is classified as Classes II and III. The 6.9 billion people on the planet now and the 7.6 billion predicted in 2020 must be fed from just 11% of the land. Because of its effects on global food security and environmental quality, land degradation is now considered one of the most important global challenges. When it comes to production, a mismatch between land usage and quality leads to land degradation (Beinroth et al., 1994). Land degradation is a biophysical process that is influenced by political and economic factors. According to a special report on climate change and land by Intergovernmental Panel on Climate Change (IPCC), land degradation is A long-term reduction or loss in the land condition that is caused by direct or indirect human-induced processes including anthropogenic activities or climate change (Johnson & Lewis, 2007).

Human activity induced land degradation threatens human welfare and jeopardizes the attainment of Sustainable Development Goals (SDGs) (Ogilvy et al., 2018). Achievement of SDGs is important for global development (Srivastava, 2023). Approximately 75% of the population of developing nations at the time was estimated by the FAO to rely on biomass for their daily energy needs (Kumar et al., 2020). Due to unsustainable logging and grazing practices, land degradation processes lower plant cover and biomass stocks, which results in net carbon emissions from soils and plant stocks (Pearson et al., 2017). In the context of climate change, the depletion of soil's organic matter pool is one of the most important chemical degradation processes. Soil organic matter pools have been reduced in agricultural soils because of the rise in respiration rates caused by tillage and the decrease in below-ground plant biomass inputs. These reductions have been further accelerated by the direct effects of warming, which affect both cultivated and uncultivated land (Bond-Lamberty et al., 2018). Degradation processes like an increase in wildfire and an increase in salinization harm the land by reducing nutrient availability and increasing metal toxicity (Minasny et al., 2017). Alterations in the hydrological system, which are crucial in the context of climate change, can also lead to the beginning of land degradation processes. Salts are primarily brought to the surface by rising water tables however, salinization led to dry to sub-humid conditions in the soil

(Schofield & Kirkby, 2003). A variety of factors, including demographic shifts, technology advancements, eating patterns and consumption habits, political, economic shifts and societal shifts, might operate as additional and more indirect drivers of land degradation (Mirzabaev et al., 2016). Temperature increases, changes in patterns and intensification of rainfall are some of the direct impacts of climate change on land degradation (Fischer & Knutti, 2015).

2.2. Overpopulation

Population growth is a key driver of development but it becomes a significant contributor to ecosystem degradation when it reaches the upper limit of the carrying capacity of the environment. According to statistics, the world's population will reach 8.6 billion people in 2030 and 9.8 billion people worldwide will populate the planet by 2050 (Valery, 2021). The renowned IPAT model (Holdren & Ehrlich, 1974) is the formal representation of this concept, in which human environmental impact (I) is thought to be a function of population size (P), per capita income (A) and technology (T) per unit of income. As the population increases, the stresses on the environment get worse every day. The environment is being threatened by the rapid population growth in developing nations like India due to the intensification and expansion of agriculture, the unplanned spread of urbanization, industrialization and the destruction of natural ecosystems (Ray & Ray, 2011). In developing countries, excessive population increase is the primary factor causing poverty, environmental damage and human suffering (Modebadze, 2021). Akinlo & Dada, (2021) noted that environmental degradation may result in high poverty levels, when necessary; resources are needed to increase human well-being and are used to address ecological concerns. The world's manufacturing bases have relocated from developed countries to developing countries due to relatively cheaper labor (Agten, 2021). Although the increase in industrial and production boosted economic activity in developing nations it also resulted in high CO_2 emissions (Kousar & Shabbir, 2021). The relationship between poverty and the environment is a complicated issue. Since the poor rely more on natural resources than the wealthy do and have little chance of accessing alternative resources, inequality may contribute to an unsustainable society (Ray & Ray, 2011). Additionally, because the poor rely heavily on natural resources, a deteriorated environment may hasten the process of poverty.

2.3. Rapid Urbanization and Industrialization

The industrialization and urbanization of a nation are important economic drivers. According to Peng and Bao, (2006), excessive use of natural resources results in increased emissions of industrial pollutants and in the end, the expansion of industry results in a decline in environmental quality. Similarly, increasing urbanization has an impact on the environment by changing the levels of polluting emissions because of an increase in production and population (Azam & Khan, 2016). Rapid urbanization increases the demand for basic infrastructure which leads to a rise in deforestation and environmental degradation (Sadorsky, 2013). Increases transportation, fuel consumption and other anthropogenic activities, leading to deuteriation of air quality (Liu & Bae, 2018), disturbs the equilibrium of natural habitats and eradicates the native species (Uttara et al., 2012). Industrialization and urbanization increase eco-environmental challenges that threatened the natural environment and sustainable development (Kavzoğlu, 2008). Unplanned urbanization led to the urban heat island phenomenon. The urban heat island phenomenon causes cities' air temperatures to rise by (1-3°C) on average during the day and by (1-2°C) at night (Bek et al., 2018).

2.4. Global Warming

Over the past few centuries, atmospheric concentrations of greenhouse gases, such as CH4, CO_2, and N2O have climbed to unprecedented levels (Usman & Makhdum, 2021). The repercussions include an increase in global temperatures, surging sea levels, deforestation, an imbalance in the climate, changes in rainfall patterns, and the spread of deserts (Cunningham et al., 1999). According to the IPCC's Mitigation of Climate Change report, "projected global emissions from country-determined contributions published before COP26 would make it likely that warming will surpass 1.5°C and also make it more difficult after 2030 to limit warming to below 2°C" (IPCC, 2022). Global agriculture is the sector responsible for 30-40% of all greenhouse emissions, making it a prominent industry that contributes to climate warming and is heavily impacted by climate change (Mishra et al., 2021). However, as a result of changing temperature and precipitation patterns, climate change has a significant impact on the growth and productivity of transboundary forests. In addition to having an impact on forest health, climate change causes specific changes in the typical structure and functions of ecosystems (Zhang et al., 2017). Climate change also has other devastating effects, including pest outbreaks, forest fires, and droughts (Abbass et al., 2022) it also threatens the livelihoods of communities that depend on forests. In agriculture, the increase in rice cultivation increases methane concentration, ruminant stocks and manure disposal increase methane, nitrous oxide and ammonia fluxes and their emission increases global warming at a higher rate (Oertel et al., 2016). Many planted species are likely to go extinct as a result of changing weather patterns brought on by a lack of natural resources (water), an increase in glacier melting, and increasing temperatures (Mihiretu et al., 2021).

2.5. Forest Degradation

In forest ecosystems, the degradation of the forest is a dominant issue. Forest degradation is generally a multi-stage process that takes place over time. It has been indicated that human activities such as fragmentation, pollution, overexploitation and fire are to be held responsible for this; however, it may also be brought on by natural events like forest fires, hurricanes, landslides, or floods (Korom et al., 2022). Forest degradation, according to the United Nations Framework Convention on Climate Change (UNFCC), is any reduction in the carbon density of forest land brought on by direct human activity (Mitchell et al., 2017). Rapid land use growth and a growing human population, particularly in developing nations, have resulted in significant alterations to the landscape that could imperil future ecological services for health and livelihood (Avtar et al., 2020). According to Gao et al., (2020), forest degradation has altered forest structure, resulting in a decline in biomass, species composition and natural regeneration. Therefore, to restore the ecosystem products and services, new ecological engineering and landscape-level methodologies are necessary to overcome the immense strain on forests brought on by anthropogenic and natural disturbances along with global climate change.

2.6. Environmental Pollution

Our diverse ecosystems are practically contaminated by anthropogenic activities, which lead to pollution. Human activities harm the environment by contaminating the water, air and soil. Clearly, in the present time urbanization and industrialization are escalating to unprecedented and alarming levels around the world. Sustainable development is threatened by pollution in developing nations (Adeuti, 2020). Because of the transboundary migration of pollutants from other nations, this inescapable issue has become extremely difficult to solve. Environmental pollution affects regular environmental processes by contaminating the physical and biological elements of the earth system to a significant degree. One reason for pollution is that the environment cannot naturally break down anthropogenically produced pollutants (Adeuti, 2020). Air pollution is the most common cause of environmental degradation. Globally, industries and automobiles are the main sources of air pollution. In addition, air pollutants that are detrimental to people include nitrogen oxide, sulfur dioxide, volatile organic compounds (VOCs), dioxins and polycyclic aromatic hydrocarbons (PAHs) (Manisalidis et al., 2020). The issue is more severe in emerging nations because of population growth, unplanned urbanization and the advancement of industrialization. This results in poor air quality, particularly in nations with social inequalities and an inadequate understanding of eco-sustainable management practices (Mannucci & Franchi, 2017).

The environment has deteriorated and polluted because of industrialization, agricultural production and urban life. This harms the rivers and oceans that are essential for life, which in turn harms human health and long-term sustainable development (Xu et al., 2022). Numerous anthropogenic activities result in the release of different contaminants into aquatic habitats. These wastes can be heavy metals, detergents, microfibers, plastic or non-plastic origin and so on, and they contribute to water pollution (Hampel et al., 2015). A significant threat to aquatic environments, human health, and productive activities is posed by the massive amounts of agricultural inputs, organic matter, sediments and saline drainage that are released into water bodies from agricultural lands (Mateo-Sagasta et al., 2017).

Soil erosion is an important concern due to the rising population at an alarming rate, increased agricultural production, development and urbanization, as well as human activity (Chen et al., 2015). The erosion process is linked to a decline in yield production, deterioration of surface water quality and impaired drainage network (Montgomery et al., 2014). Rainfall is one of the main contributors to soil erosion because it loosens the soil, erodes it and then washes it away as runoff. The changes in the land use type have an impact on the process of soil erosion as well. Water and wind are the two main factors that contribute to large amounts of yearly soil loss (Issaka and Ashraf, 2017).

3. Effect of Ecosystem Degradation

3.1. Human Health

An integrated strategy is needed to address issues of development, environment and human health. Human health appears to be drastically affected by degraded aquatic and terrestrial ecosystems. Odum (1995) characterized the deterioration of human health by a disrupted ecosystem as an "illness resulting from the disturbed internal balance due to external disturbances." Population health is affected in many direct and indirect ways by the disturbed ecosystem. There are several negative health effects of air pollution. According to estimates, air pollution causes more than 30% of ischemic heart disease and stroke fatalities, 17% of deaths from chronic obstructive pulmonary disease, 9% of deaths from respiratory infections and 9% of deaths from lung cancer globally (Mannucci & Franchini, 2017). Numerous studies have demonstrated that both short- and long-term exposure to particulate matter (PM) is linked to an increase in mortality. After inhalation, PM10 particles can penetrate the lungs and even enter the bloodstream. PM2.5, or fine particles, present a larger threat to health (Zhang et al., 2019). Additionally, it was found that long-term exposure to PM was linked to both newborn mortality and cardiovascular disorders. It is important to note that those with conditions like asthma, pneumonia, diabetes, and cardiovascular and respiratory illnesses are particularly prone to the effects of PM (Manisalidis et al., 2020).

Numerous human activities result in the release of different pollutants into aquatic ecosystems, endangering the health of living things and degrading the environment by making water bodies uninhabitable (Bashir et al., 2020). According to the UNESCO 2021 World Water Development Report, almost three lakh children under the age of five (5.3% of all deaths in this age group) die from diarrhea every year because of the consumption of contaminated water, poor sanitation and poor hygiene practices. Poverty and lack of access to water and sanitary amenities also raise the prevalence of illnesses like cholera, trachoma, schistosomiasis and helminthiasis (Lin et al., 2022). Enhance public awareness through the newspaper, television and other media, introducing environmental education and providing training to farmers to avoid overuse of chemicals to reduce contamination of water and impact on human health.

3.2. Ozone Layer Depletion

Ozone (O3), a stratospheric layer, is crucial for sustaining humans and ensuring their survival. This layer blocks 93-99% of the sun's high-frequency UV light, which may be harmful to earthly life (Albritton, 1998). This layer blocks the ultraviolet rays which cause cell damage. With the development of the ozone layer comes the emergence of complex living forms. Normal ozone concentration is 300-350 Dobson units but when the concentration drops below 200 Dobson units it is called ozone layer depletion. Intermittent ozone loss can result from natural events, but it is widely recognized that man-made substances like CFCs which produce chlorine and bromine are the primary source of this depletion (Angella & Korshover, 1973). UV-B radiation increases the risk of eye diseases like damage to the cornea or cataract, lead to skin cancer and alter the immune system causing several infectious diseases. Increased UV-B radiation can bring changes in forests and grasslands by bringing changes in the species composition due to mutation, which would also alter the biodiversity in various ecosystems (Sivasakthivel & Reddy, 2011). Increased UV-B radiation also has other effects on the growth of phytoplankton in the upper ocean, changes in the production and decomposition of plant matter and decreased primary production (Brenna et al., 2019).

3.3. Loss of Biodiversity

Biodiversity is crucial for human well-being. However, one of the greatest environmental threats to humanity is the loss of ecosystem functions and services. Many ecosystem processes, such as community structure and interactions, are disrupted by biodiversity loss and this can lead to ecosystem breakdowns ranging from decreasing biomass productivity to lowering ecosystem elasticity (Lu et al., 2020). Habitat modification, overexploitation, invasive alien species, climate change and chain extinctions are the main causes of the current loss of biodiversity (Sonwa et al., 2017). Climate change is one of the major threats to biodiversity and ecosystem services

(Matata & Adan, 2018). Effects of climate change are expected to be the primary forces influencing biodiversity at all levels, along with anthropogenic impacts (Parmesan, 2006). In the next 50 years, climate change might lead to the extinction of more than a million terrestrial species, according to a paper published in Nature (Thomas et al., 2004). The hydrological cycle can modify biodiversity in a variety of ways by shifting species distribution and habitat ranges, altering life cycles, affecting migration patterns, and the frequency and intensity of pest and disease outbreaks (Sintayehu, 2018). Ecosystem services affect 40% of the world's economy and 80% of the economy of developing nations (Travis, 2003). Significant efforts must be made in addressing the causes of biodiversity loss. Future efforts must begin to address the gap between the relevant sectors. Efforts can be made by increasing knowledge, awareness and responsibility toward changing conditions.

3.4. Desertification

Desertification is an irretrievable form of land degradation. Overgrazing, secondary salinization and other processes have a detrimental effect on the ecosystem services in drylands (Majeed & Muhammad, 2019). As vegetation cover decreases due to desertification, the natural soil alters, disrupting the albedo and water balance (Gonzalez-Martin et al., 2014). Some plant species are already in danger of becoming extinct, especially those with limited habitats and rising aridity makes that risk even worse (Gitay et al., 2002). Desertification due to land use changes also leads to the loss of biodiversity in the dryland area (Wilting et al., 2017). Increasing climate change will make some dryland areas even more parched and increase the frequency of droughts. Mammals and birds may be especially vulnerable to droughts because they depend on evaporative cooling to retain their internal temperature within a normal range and face the possibility of fatal dehydration in habitats with limited water supplies (Albright et al., 2017). Desertification led to the loss of agricultural productivity and income. Desertification changes the composition of species with low forage quality and reduces the productivity of livestock and increases livestock diseases (Dibari et al., 2016). Cultivation of wheat, maize and rice with unsustainable land management leads to a global economic loss of 56.6 billion USD due to lower livestock productivity which is caused due to rangeland degradation (Nkonya et al., 2016). It is important to have a better understanding of the reason that led to desertification. Further research and development of technology are required to combat desertification.

Conclusion

Ecosystem degradation is one of the most serious problems faced by humankind. City planners, industry leaders and managers of natural resources must take the environment's long-term consequences into account to lessen future repercussions. It will take multiple efforts to successfully reduce environmental degradation and restore our ecology. The deterioration of our ecosystem can be slowed down in several ways. The government must prioritize filling in the gaps in the legal system to prevent illegal activity because this is the most efficient way to reduce pollution. All levels of authority must be involved in the planning of policies, their execution and the monitoring of progress with a few sustainable steps and a switch to a green economy; we can prevent the deterioration of our ecosystem. According to the study, policymakers should establish technology-friendly environments that can help to overcome pollution, to prevent land and forest degradation which can help to improve the environment in a better way. Further, increasing environmental awareness among the public will help us to create a sustainable environment.

References

Abbass, K., Qasim, M. Z., Song, H., Murshed, M., Mahmood, H., and Younis, I. (2022). A review of the global climate change impacts, adaptation, and sustainable mitigation measures. *Environmental science and pollution research international*, 29(28), 42539–42559. https://doi.org/10.1007/s11356-022-19718-6.

Adeuti, D. B. R. (2020). Analysis of environmental pollution in developing countries. *American* Academic *Scientific Research Journal for Engineering*, 65(1), 39–48.

Agten, S. (2021). 'Made in China' No longer means cheap. In *Adventures in the Chinese economy: 16 years from the inside* (pp. 1–18). Palgrave Macmillan.

Akinlo, T., and Dada, J. T. (2021). The moderating effect of foreign direct investment on environmental degradation-poverty reduction nexus: Evidence from sub-Saharan African countries. *Environment, Development and Sustainability*, 23(11), 15764–15784. https://doi.org/10.1007/s10668-021-01315-1.

Albright, T. P., Mutiibwa, D., Gerson, A. R., Smith, E. K., Talbot, W. A., O'Neill, J. J., McKechnie, A. E., and Wolf, B. O. (2017). Mapping evaporative water loss in desert passerines reveals an expanding threat of lethal dehydration. *Proceedings of the National Academy of Sciences of the United States of America*, 114(9), 2283–2288. https://doi.org/10.1073/pnas.1613625114.

Albritton, D. (1998). *What should be done in a science assessment in protecting the ozone layer: Lessons, models, and prospects*.

Angell, J. K., and Korshover, J. (1973). Quasi-biennial and long-term fluctuations in total ozone. *Monthly Weather Review*, 101(5), 426–443. https://doi.org/10.1175/1520-0493(1973)101<0426:QALFIT>2.3.CO;2.

Avtar, R., Komolafe, A. A., Kouser, A., Singh, D., Yunus, A. P., Dou, J., Kumar, P., Das Gupta, R. D., Johnson, B. A., Thu Minh, H. V., Aggarwal, A. K., and Kurniawan, T. A. (2020). Assessing sustainable development prospects through remote sensing: A review. *Remote Sensing Applications: Society and Environment*, 20, 100402. https://doi.org/10.1016/j.rsase.2020.100402.

Azam, M., and Khan, A. Q. (2016). Urbanization and environmental degradation: Evidence from four SAARC countries Bangladesh, India, Pakistan, and Sri Lanka. *Environmental Progress and Sustainable Energy*, 35(3), 823–832. https://doi.org/10.1002/ep.12282.

Bashir, I., Lone, F. A., Bhat, R. A., Mir, S. A., Dar, Z. A., and Dar, S. A. (2020). Concerns and Threats of Contamination on Aquatic Ecosystems. *Bioremediation and Biotechnology: Sustainable Approaches to Pollution Degradation*, 1–26. https://doi.org/10.1007/978-3-030-35691-0_1.

Beinroth, F. H., Eswaran, H., Reich, P. F. and Van den berg, E. (1994). Land related stresses in agroecosystems. In: *Stressed Ecosystems and Sustainable Agriculture*, eds. S.M. Virmani, J.C. Katyal, H. Eswaran, and I.P. Abrol. New Delhi: Oxford and IBH.

Bek, M. A., Azmy, N., and Elkafrawy, S. (2018). The effect of unplanned growth of urban areas on heat island phenomena. *Ain Shams Engineering Journal*, 9(4), 3169–3177. https://doi.org/10.1016/j.asej.2017.11.001.

Bond-Lamberty, B., Bailey, V. L., Chen, M., Gough, C. M., and Vargas, R. (2018). Globally rising soil heterotrophic respiration over recent decades. *Nature*, 560(7716), 80–83. https://doi.org/10.1038/s41586-018-0358-x.

Brenna, H., Kutterolf, S., and Krüger, K. (2019). Global ozone depletion and increase of UV radiation caused by pre-industrial tropical volcanic eruptions. *Scientific Reports*, 9(1), 9435. https://doi.org/10.1038/s41598-019-45630-0.

Brown, B. J., Hanson, M. E., Liverman, D. M., and Merideth, R. W. (1987). Global sustainability: Toward definition. *Environmental Management*, 11(6), 713–719. https://doi.org/10.1007/BF01867238.

Burki, M. A. K., Burki, U., and Najam, U. (2021). Environmental degradation and poverty: A bibliometric review. *Regional Sustainability*, 2(4), 324–336. https://doi.org/10.1016/j.regsus.2022.01.001.

Cao, Y., Kong, Lingqiao, Zhang, Lufeng, and Ouyang, Z. (2020). Spatial characteristics of ecological degradation and restoration in China from 2000 to 2015 using remote sensing. *Restoration Ecology*, 28(6), 1419–1430. https://doi.org/10.1111/rec.13226.

Chakravarty, D., and Mandal, S. K. (2020). Is economic growth a cause or cure for environmental degradation? Empirical evidences from selected developing economies. *Environmental and Sustainability Indicators*, 7, 100045. https://doi.org/10.1016/j.indic.2020.100045.

Chen, K., Wang, X., Li, D., and Li, Z. (2015). Driving force of the morphological change of the urban lake ecosystem: A case study of Wuhan, 1990–2013. *Ecological Modelling*, 318, 204–209. https://doi.org/10.1016/j.ecolmodel.2015.06.017.

Cunningham, W. P., Cooper, T. H., Gorham, E., and Hepworth, M. T. (1998). *Environmental encyclopedia* (pp. 1–110).

Delgado, L. E., and Marín, V. H. (2020). Ecosystem services and ecosystem degradation: Environmentalist's expectation? *Ecosystem Services*, 45. https://doi.org/10.1016/j.ecoser.2020.101177.

Dibari, C., Bindi, M., Moriondo, M., Staglianò, N., Targetti, S., and Argenti, G. (2016). Spatial data integration for the environmental characterization of pasture macrotypes in the Italian Alps. *Grass and Forage Science*, 71(2), 219–234. https://doi.org/10.1111/gfs.12168.

Fischer, E. M., and Knutti, R. (2015). Anthropogenic contribution to global occurrence of heavy-precipitation and high-temperature extremes. *Nature Climate Change*, 5(6), 560–564. https://doi.org/10.1038/nclimate2617.

Gao, Y., Skutsch, M., Paneque-Gálvez, J., and Ghilardi, A. (2020). Remote sensing of forest degradation: A review. *Environmental Research Letters*, 15(10). https://doi.org/10.1088/1748-9326/abaad7.

Gitay, H., Suárez, A., Dokken, D. J., and Watson, R. T. (2002) Climate Change and Biodiversity. *Intergovernmental Panel on Climate Change Technical Paper V*.

Gonzalez-Martin, C., Teigell-Perez, N., Valladares, B., and Griffin, D. W. (2014). The Global Dispersion of Pathogenic Microorganisms by Dust Storms and Its Relevance to Agriculture. *Advances in Agronomy*, 127, 1–41. https://doi.org/10.1016/B978-0-12-800131-8.00001-7.

Gordon, L. J., and Enfors, E. I. (2008). "*Land degradation, ecosystem services and resilience of smallholder farmers in Makanya Catchment, Tanzania*," IWMI Books, Reports H041592, International Water Management Institute.

Hampel, M., Blasco, J., and Segner, H. (2015). Molecular and cellular effects of contamination in aquatic ecosystems. *Environmental Science and Pollution Research International*, 22(22), 17261–17266. https://doi.org/10.1007/s11356-015-5565-5.

Holdren, J. P., and Ehrlich, P. R. (1974). Human population and the global environment. *American Scientist*, 62(3), 282–292.

Intergovernmental Panel on Climate Change. (2022). 2022: Impacts, adaptation, and vulnerability. *Working Group II contribution to the sixth assessment report of the Intergovernmental Panel on Climate Change*. https://www.ipcc.ch/report/ar6/wg2/downloads/report/IPCC_AR6_WGII_FinalDraft_FullReport.pdf. *Climate Change*. Geneva.

Issaka, S., and Ashraf, M. A. (2017). Impact of soil erosion and degradation on water quality: A review. *Geology, Ecology, and Landscapes*, 1(1), 1–11. https://doi.org/10.1080/24749508.2017.1301053.

Johnson, D. L., and Lewis, L. A. (2007). *Land degradation: Creation and destruction*. Rowman & Littlefield Publishing Group.

Kavzoğlu, T. (2008). Determination of environmental degradation due to urbanization and industrialization in Gebze, Turkey. *Environmental Engineering Science*, 25(3), 429–438. https://doi.org/10.1089/ees.2006.0271.

Korom, A., Jalloh, M. B., Jamian, M. Azizul H., and Latif, Z. A. (2022). Understanding forest degradation: A review of forest structue indicators. *Borneo akademika* [*Borneo Academy*]. 6(1), 115-125.

Kousar, S., and Shabbir, A. (2021). Analysis of environmental degradation mechanism in the nexus among energy consumption and poverty in Pakistan. *Environmental Science and Pollution Research International*, 28(22), 27528–27541. https://doi.org/10.1007/s11356-020-12140-w.

Kumar, V., Kumar, P., and Khan, A. (2020). Optimization of PGPR and silicon fertilization using response surface methodology for enhanced growth, yield and biochemical parameters of French bean (*Phaseolus vulgaris L.*) under saline stress. *Biocatalysis and Agricultural Biotechnology*, 23, 101463. https://doi.org/10.1016/j.bcab.2019.101463.

Lin, L., Yang, H., and Xu, X. (2022). Effects of water pollution on human health and disease heterogeneity: A review. *Frontiers in Environmental Science*, 10. https://doi.org/10.3389/fenvs.2022.880246.

Liu, X., and Bae, J. (2018). Urbanization and industrialization impact of CO_2 emissions in China. *Journal of Cleaner Production*, 172, 178–186. https://doi.org/10.1016/j.jclepro.2017.10.156.

Lu, Y., Yang, Y., Sun, B., Yuan, J., Minzhao, Y., Stenseth, N. Chr., Bullock, J. M., and Obersteiner, M. (2020). *"APPLIED ECOLOGY Spatial Variation in Biodiversity Loss Across China Under Multiple Environmental Stressors*. http://www.nsii.org.cn/.

Majeed, A., and Muhammad, Z. (2019). Salinity: A major agricultural problem – Causes, impacts on crop productivity and management strategies. In M. Hasanuzzaman, K. R. Hakeem, K. Nahar & H. Alharby (Eds.), *Plant abiotic stress tolerance* (pp. 83–99). Springer International Publishing.

Manisalidis, I., Stavropoulou, E., Stavropoulos, A., and Bezirtzoglou, E. (2020). Environmental and Health Impacts of Air Pollution: A Review. *Frontiers in public health*, 8, 14. https://doi.org/10.3389/fpubh.2020.00014.

Mannucci, P. M., and Franchini, M. (2017). Health effects of ambient air pollution in developing countries. *International Journal of Environmental Research and Public Health. MDPI*, 14(9). https://doi.org/10.3390/ijerph14091048.

Matata, A. C., and Adan, A. (2018). Causes of climate change and its impact in the multi-sectoral areas in Africa-Need for enhanced adaptation policies. *Current Journal of Applied Science and Technology*, *27*(3), 1–10. https://doi.org/10.9734/CJAST/2018/37276.

Mateo-Sagasta, J., Zadeh, S. M., Turral, H., and Burke, J. (2017). *Water pollution from agriculture: A global review.* Food and Agriculture Organization of the United Nations, Rome and the International Water Management Institute on behalf of the Water Land and Ecosystems Research Program, Colombo.

Maurya, P. K., Ali, S. A., Ahmad, A., Zhou, Q., da Silva Castro, J., Khane, E., and Ali, H. (2020). An introduction to environmental degradation: Causes, consequence and mitigation. In V. Kumar, J. Singh & P. Kumar (Eds.), *Environmental degradation:*

Causes and remediation strategies, 1 (pp. 1–20). https://doi.org/10.26832/aesa-2020-edcrs-01.

Mihiretu, A., Okoyo, E. N., and Lemma, T. (2021). Awareness of climate change and its associated risks jointly explain context-specifc adaptation in the Arid-tropics, Northeast Ethiopia. *n SN Social Sciences*, 1(2), 1–51.

Millennium Ecosystem Assessment. (2005). *Ecosystems and human well-being: Synthesis*. World Resources Institute.

Minasny, B., Malone, B. P., Mcbratney, A. B., Angers, D. A., Arrouays, D., Chambers, A., Chaplot, V., Chen, Z. S., Cheng, K., Das, B. S., Field, D. J., Gimona, A., Hedley, C. B., Hong S. Y., Mandal, B., Marchant, B. P., Martin, M., McConkey, B. G., Mulder, V. L., O'Rourke, S., Richer-de-Forges, A. C., Odeh, I., Padarian, J., Paustian, K., Pan, G., Poggio, L., Savin, I., Stolbovoy, V., Stockmann, U., Sulaeman, Y., Tsui, C. C., Vågen, T. G., van Wesemael, B., and Winowiecki, L. (2017). Soil carbon 4 per mille. *Geoderma.*, 292, 59–86.

Mirzabaev, A., Nkonya, E., Goedecke, J., Johnson, T., and Anderson, W. (2016). *Global drivers of land degradation and improvement. Economics of land degradation and improvement – A global assessment for sustainable development* (pp. 167–195). Springer International Publishing.

Mishra, A., Bruno, E., and Zilberman, D. (2021). Compound natural and human disasters: Managing drought and COVID-19 to sustain global agriculture and food sectors. *Science of the Total Environment*, *754*, 142210. https://doi.org/10.1016/j.scitotenv.2020.142210.

Mitchell, A. L., Rosenqvist, A., and Mora, B. (2017). Current remote sensing approaches to monitoring forest degradation in support of countries measurement, reporting and verification (MRV) systems for REDD. *Carbon Balance and Management*, *12*(1), 9. https://doi.org/10.1186/s13021-017-0078-9.

Modebadze, V. (2021). (2021). The issue of unchecked population growth and its effect on environmental degradation. *KutBilim Sosyal Bilimler ve Sanat Dergisi [KutBilim Social Sciences and Art Journal]*, 1 (1), 36-44. Retrieved from https://dergipark.org.tr/en/pub/kutbilim/issue/69572/1109621.

Montgomery, D. R., Huang, M. Y. F., and Huang, A. Y. L. (2014). Regional soil erosion in response to land use and increased typhoon frequency and intensity, Taiwan. *Quaternary Research*, 81(1), 15–20. https://doi.org/10.1016/j.yqres.2013.10.005.

Nkonya, E., Mirzabaev, A., and von Braun, J. (2016). Global cost of land degradation. In E. Nkonya, A. Mirzabaev & J. von Braun (Eds.), *Economics of land degradation and improvement – à global assessment for sustainable development* (pp. 117–165). Springer International Publishing.

O'Neill, B. C., Dalton, M., Fuchs, R., Jiang, L., Pachauri, S., and Zigova, K. (2010). Global demographic trends and future carbon emissions. *Proceedings of the National Academy of Sciences of the United States of America*, 107(41), 17521–17526. https://doi.org/10.1073/pnas.1004581107.

Odum, E. P. (1995). Profile analysis and some thoughts on the development of the interface area of environmental health. *Ecosystem Health*, 1, 41–45.

Oertel, C., Matschullat, J., Zurba, K., Zimmermann, F., and Erasmi, S. (2016). Greenhouse gas emissions from soils – A review. *Geochemistry*, 76(3), 327–352. https://doi.org/10.1016/J.CHEMER.2016.04.002.

Ogilvy, S., Burritt, R., Walsh, D., Obst, C., Meadows, P., Muradzikwa, P., and Eigenraam, M. (2018). Accounting for liabilities related to ecosystem degradation. *Ecosystem Health and Sustainability*, 4(11), 261–276. https://doi.org/10.1080/20964129.2018.1544837.

Pan, Z., He, J., Liu, D., Wang, J., and Guo, X. (2021). Ecosystem health assessment based on ecological integrity and ecosystem services demand in the Middle Reaches of the Yangtze River Economic Belt, China. *Science of the Total Environment*, 774, 144837. https://doi.org/10.1016/j.scitotenv.2020.144837.

Parmesan, C. (2006). Ecological and evolutionary responses to recent climate change. *Annual Review of Ecology, Evolution, and Systematics*, 37(1), 637–669. https://doi.org/10.1146/annurev.ecolsys.37.091305.110100.

Pearson, T. R. H., Brown, S., Murray, L., and Sidman, G. (2017). Greenhouse gas emissions from tropical forest degradation: An underestimated source. *Carbon Balance and Management*, 12(1), 3. https://doi.org/10.1186/s13021-017-0072-2.

Peng, S., and Bao, Q. (2006). Economic growth and environmental pollution: An empirical test for the environmental Kuznets Curve hypothesis in China. *Research on Financial and Economic Issues*, 273(8), 3–17.

Ray, S., and Ray, I. (2011). Impact of Population Growth on Environmental Degradation: Case of India. *Journal of economics and sustainable development, 2*, 72-77.

Sadorsky, P. (2013). Do urbanization and industrialization affect energy intensity in developing countries? *Energy Economics*, 37, 52–59. https://doi.org/10.1016/j.eneco.2013.01.009.

Schofield, R. V., and Kirkby, M. J. (2003). Application of salinization indicators and initial development of potential global soil salinization scenario under climatic change. *Global Biogeochemical Cycles*, 17(3), n/a–n/a. https://doi.org/10.1029/2002GB001935.

Sintayehu, D. W. (2018). Impact of climate change on biodiversity and associated key ecosystem services in Africa: A systematic review. *Ecosystem Health and Sustainability*, 4(9), 225–239. https://doi.org/10.1080/20964129.2018.1530054.

Sivasakthivel, T., and Reddy, K. K. S. K. (2011). Ozone layer depletion and its effects: A review *International Journal of Environmental Science and Development*, 2(1), 30–37. https://doi.org/10.7763/ijesd.2011.v2.93.

Sonwa, D. J., Dieye, A., El Mzouri, E. H., Majule, A., Mugabe, F. T., Omolo, N., Wouapi, H., Obando, J., and Brooks, N. (2017). Drivers of climate risk in African agriculture. *Climate and Development*, 9(5), 383–398. https://doi.org/10.1080/17565529.2016.1167659.

Srivastava, R. (2023). *Strategies to Achieve Sustainable Development Goals (SDGs): A Road Map for Global Development*. Nova Science Publishers https://doi.org/10.52305/YNDL2610.

Sun, Q., Qi, W., and Yu, X. (2021). Impacts of land use change on ecosystem services in the intensive agricultural area of North China based on multi-scenario analysis.

Alexandria Engineering Journal, 60(1), 1703–1716. https://doi.org/10.1016/j.aej.2020.11.020.

Thomas, C. D., Cameron, A., Green, R. E., Bakkenes, M., Beaumont, L. J., Collingham, Y. C., Erasmus, B. F. N., De Siqueira, M. F., Grainger, A., Hannah, L., Hughes, L., Huntley, B., Van Jaarsveld, A. S., Midgley, G. F., Miles, L., Ortega-Huerta, M. A., Peterson, A. T., Phillips, O. L., and Williams, S. E. (2004). Extinction risk from climate change. *Nature*, 427(6970), 145–148. https://doi.org/10.1038/nature02121.

Travis, J. M. J. (2003). Climate change and habitat destruction: A deadly anthropogenic cocktail. *Proceedings. Biological Sciences. Proceedings of the R Social Lond Series*, 270(1514), 467–473. https://doi.org/10.1098/rspb.2002.2246.

Usman, M., and Makhdum, M. S. A. (2021). What abates ecological footprint in BRICS-T region? Exploring the influence of renewable energy, non-renewable energy, agriculture, forest area and financial development. *Renewable Energy*, 179, 12–28. https://doi.org/10.1016/j.renene.2021.07.014.

Uttara S., Bhuvandas N., and Aggarwal V. (2012). Impacts of urbanization on Environment. *International Journal of Research in Engineering & Applied Sciences*, 2 (2), 1637-1645.

Wilting, H. C., Schipper, A. M., Bakkenes, M., Meijer, J. R., and Huijbregts, M. A. J. (2017). Quantifying biodiversity losses due to human consumption: A global-scale footprint analysis. *Environmental Science and Technology*, 51(6), 3298–3306. https://doi.org/10.1021/acs.est.6b05296.

Xu, X., Yang, H., and Li, C. (2022a). Theoretical model and actual characteristics of air pollution affecting health cost: A review. *International Journal of Environmental Research and Public Health*, 19(6), 3532. https://doi.org/10.3390/ijerph19063532.

Zhang, L., Yang, Y., Li, Y., Qian, Z. (M.), Xiao, W., Wang, X., Rolling, C. A., Liu, E., Xiao, J., Zeng, W., Liu, T., Li, X., Yao, Z., Wang, H., Ma, W., and Lin, H. (2019). Short-term and long-term effects of PM2.5 on acute nasopharyngitis in 10 communities of Guangdong, China. *Science of the Total Environment*, 688, 136–142. https://doi.org/10.1016/j.scitotenv.2019.05.470.

Zhang, M., Liu, N., Harper, R., Li, Q., Liu, K., Wei, X., Ning, D., Hou, Y., and Liu, S. (2017). A global review on hydrological responses to forest change across multiple spatial scales: Importance of scale, climate, forest type and hydrological regime. *Journal of Hydrology*, 546, 44–59. https://doi.org/10.1016/j.jhydrol.2016.12.040.

Chapter 3

The Complexity of Ecological Networks: A Computational Analysis Using R

Basant K. Tiwary*
Department of Bioinformatics, School of Life Sciences, Pondicherry University, Pondicherry, India

Abstract

An ecosystem consists of systems-level interactions of interlinked species. A holistic model of an ecosystem is represented in the form of a network model consisting of species as nodes and interactions among them as edges. The positive and negative interaction between species may be realized in the form of mutualism and predation/parasitism respectively. The host-parasite interaction and plant-pollinator interaction are computationally modeled using a bipartite network. This model helps us in understanding the overall stability and controllability of an ecosystem. The network complexity is measured using some common measures such as species richness, connectance, connectivity, linkage density and node degree distribution. The network architecture is also measured regarding modularity, nestedness and motifs. Network robustness is usually studied at the node level and edge level to understand the stability of an ecosystem. There are various databases available for ecological interactions across different species in diverse ecosystems. R is an open-source language for statistical computing providing more than 18,000 packages for computational analysis. Several R packages are available for access to interaction databases, computational analysis and visualization of ecological networks. Here, I have demonstrated two hypothetical examples of reconstruction and

* Corresponding Author's Email: basant68@gmail.com.

In: Conserving, Halting and Reinstating Degraded Natural Resources …
Editors: Rajani Srivastava and Pankaj Kumar
ISBN: 979-8-89113-067-8

analysis of ecological networks. The first network is a simple food web network reconstructed and analyzed using a popular R package igraph. The second network is a bipartite network consisting of hypothetical data of plant-pollinator interactions that are analyzed using two R packages igraph and bipartite. These computational analyses of ecological networks may provide useful insights into the inherent complexity of an ecosystem.

Keywords: ecological network, network complexity, databases, R language, R packages

1. Introduction

Ecological systems are nonlinear dynamic systems consisting of a combination of agonistic and mutualistic interactions between species (Pascual et al., 2006). The ecosystem consists of a series of interlinked species either interacting directly or through environmental factors such as space, light, and environment (Montoya et al., 2006). Therefore, a systems-level understanding of the ecosystem is necessary to conserve the existing biodiversity. The network models of the ecosystem are useful not only in understanding the inherent complexity of a natural ecosystem but also in understanding perturbed ecosystems subjected to stress as well. An ecological network model consists of a set of compartments, usually indicating a species and links showing either species-to-species/species-to-environment interactions or biomass/energy flows across the compartments (Zhao et al., 2018). This network can be perturbed either using biotic components or abiotic components. The perturbation of biotic components can be realized by the extinction of a species or the introduction of a new species or the decline in the abundance of a species. Similarly, abiotic components can be perturbed by either change in climate or fragmentation of habitat (Hooper et al., 2005). The connection between the complexity and stability of an ecosystem is equally challenging for both computational biologists and ecologists. The stability and controllability of an ecosystem are tightly regulated by the interactions among key nodes in the network (Feng & Takemoto, 2014). The presence of some crucial species is necessary to maintain ecological stability. Moreover, the diversity of species and their interactions lead to ecosystem stability (Cagua et al., 2019). Thus, ecological network models are necessary for understanding ecosystem complexity and stability.

2. Ecological Networks

Ecological networks consist of multiple types of interactions among species inhabiting a particular community such as trophic interactions, mutualistic interactions, or competitive interactions (Mougi & Kondoh, 2012). An ecological network has a set of N nodes representing different individuals/species/functional groups of species and L links connecting the nodes representing interactions across each pair of species (Montoya et al., 2006). These internode interactions may be competition, predation and mutualism in an ecological network (Chesson & Kuang, 2008). These links can be either weighted links showing the strength of interaction or unweighted links described by zero (absence of a link) and one (presence of a link). When both interacting species have an equal effect on each other, the link is undirected. By contrast, one species may have a differential effect on other species and this interaction is represented by a directed link (van Altena et al., 2016). The interaction between two species is also indicated by a positive or negative coefficient. For example, in trophic networks based on food webs, because predator species feed on prey species, the effect of a predator on prey is a negative coefficient whereas the effect of prey on the predator is a positive coefficient. The coefficients for both interacting species are positive in mutualistic networks but these coefficients are negative in competitive networks (Ndour et al., 1996).

An ecological network is defined by a matrix, where each row and column represent interacting species. Each element of the matrix represents the link between two species. An undirected and unweighted network matrix is symmetric and elements are either one or zero (Figure 1). Moreover, the undirected and weighted network matrix has real values (Figure 2). By contrast, a directed and unweighted network has elements that are either -1 (negative) or +1 (positive) or 0 (neutral) (Figure 3) but each element of the matrix is also a real value for a directed and weighted network (Estrada, 2012; Newman, 2010;) (Figure 4). The interaction between two disjoint sets of species such as host-parasite interaction and plant-pollinator are represented by a bipartite (two-mode) network (Dormann et al., 2009).

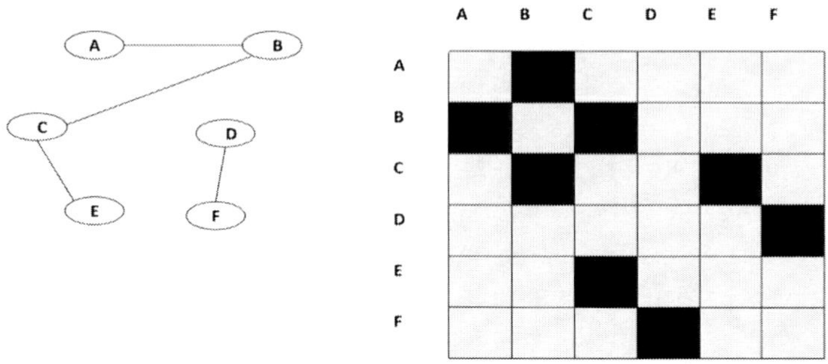

Figure 1. Unweighted undirected ecological network and its corresponding matrix showing the presence (black) and absence (grey) of interactions.

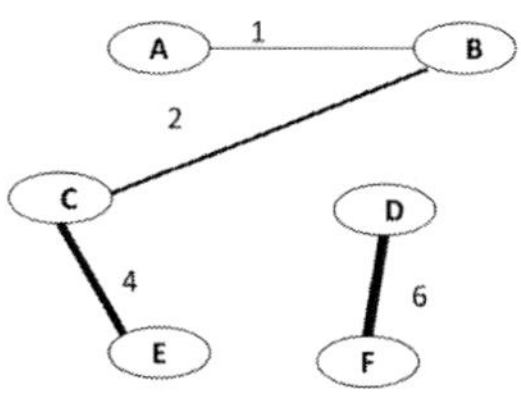

	A	B	C	D	E	F
A	0	1	0	0	0	0
B	1	0	2	0	0	0
C	0	2	0	0	4	0
D	0	0	0	0	0	6
E	0	0	4	0	0	0
F	0	0	0	6	0	0

Figure 2. Weighted undirected ecological network of six species and its corresponding matrix showing the real values of interactions between species.

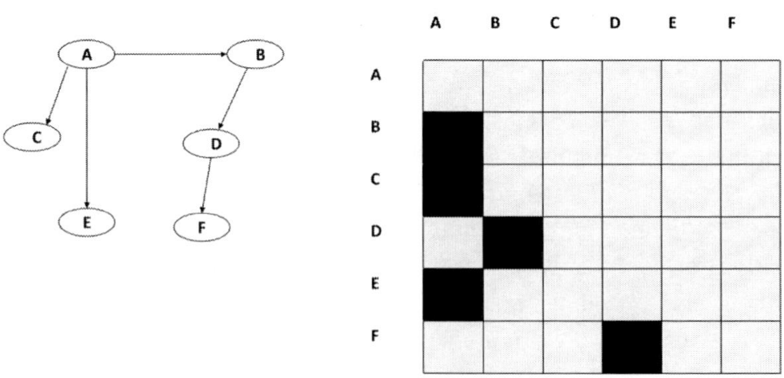

Figure 3. Unweighted directed ecological network and its corresponding matrix showing the presence (black) and absence (grey) of interactions.

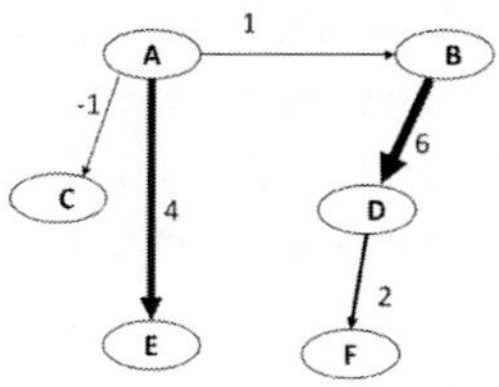

	A	B	C	D	E	F
A	0	0	0	0	0	0
B	1	0	0	0	0	0
C	-1	0	0	0	0	0
D	0	6	0	0	0	0
E	4	0	0	0	0	0
F	0	0	0	2	0	0

Figure 4. Weighted directed ecological network and its corresponding matrix showing real values of interactions. Species A has a positive effect on species B and a negative effect on species C.

In contrast, interactions among species in a food web constitute a unipartite (one-mode) network. There are various approaches to the quantification of elements in the matrix. The first approach is to treat each element of a community matrix as an effect of perturbation from the equilibrium of the abundance of one species on the population growth rate of another species (Novak et al., 2016). The second approach is to describe the elements as interaction strengths of an Interaction matrix (Novak et al., 2016; Kokkoris et al., 2002). However, these two approaches are suitable for theoretical studies but difficult to measure in an empirical or field study. Empirical studies are relatively easy to measure and are largely based on the amount of energy flow and biomass flow in an ecosystem or prey-predator interaction (Jordano et al., 2002). However, there is no relationship between these values and elements of a theoretical community matrix as measured values are not reliant on species equilibrium abundances (Dunne et al., 2002a).

2.1. Measures of Network Complexity

The complexity of an ecological network can be measured by some descriptors which are as follows.

2.1.1. Species Richness (S)

Species richness is the total number of interacting species in the network (i.e., network size) (MacArthur, 1955; May 1972). Here, we focus on trophic species (i.e., functional groups of species sharing the same trophic level) rather than taxonomic species (i.e., species having a morphological and phylogenetic relationship) in food web studies (Pimm et al., 1991; Schoener 1989). In some cases, morphological similarity among species is considered a criterion to judge morphospecies in the absence of distinct taxonomic differences (Olito & Fox, 2015).

2.1.2. Connectance (C)

It measures the proportion of realized interactions out of all possible interactions by all species in an ecological network (L/S^2) (Estrada, 2012; May 2019).

2.1.3. Connectivity (L)

This measure is a simpler measure than connectance. It measures only the total number of interactions of all species present in a network (Newman, 2010; Okuyama & Holland, 2008).

2.1.4. Linkage Density (L/S)

Linkage density is computed as the average number of links per species or connectivity divided by species richness (Montoya et al., 2006; Pimm et al., 1991). It helps us in discriminating between generalist species (nodes having many interactions) and specialist species (nodes having few interactions) in a food-web network (Havens, 1992).

2.1.5. Node Degree Distribution

The node degree is the total number of interactions of a species in an ecological network (Newman 2010). This measure emphasizes the relative role of each species in an ecological network. It is represented as the distribution of the number of interactions per species (Camacho et al., 2002).

2.2. Measures of Ecological Network Architecture

An ecological network is characterized by certain topological features. Two important measures of architecture in an ecological network are modularity and nestedness which are described as follows:

2.2.1. Modularity

The modularity of a network measures the strength of the division of an ecological network into modules (States, 2004). The species in a particular module are likely to interact strongly within the module but they have weak interactions with species present in another module (Olesen et al., 2007).

2.2.2. Nestedness

This measure is important for mutualistic ecological networks. Here, both generalist species and specialist species prefer to interact with only generalist species in a nested network (Bascompte et al., 2003). However, specialist species tend to avoid interaction with other specialist species. The quantitative measures of nestedness in an ecological network are temperature and nestedness metrics based on Overlap and Decreasing fill (NODF) (Almeida-Neto et al., 2008; Bascompte et al., 2003).

2.2.3. Motifs

Motifs are some patterns observed frequently in real ecological networks in comparison to an ensemble of randomized networks (Milo et al., 2002; Simmons et al., 2019). In other words, these are subnetworks or building blocks showing interactions among a specific group of species. A motif is formed from a minimal number of two species and is useful in determining the role of a species in a network. Motifs characterize an ecological network efficiently in comparison to other indices such as species degree and nestedness and are more insensitive to sampling effects (Delmas et al., 2019). Identical index values can be obtained with ecological networks having distinct configurations. Moreover, indirect interactions are completely ignored in these one-dimensional indices. Motifs are useful in deciphering indirect interactions in ecological communities such as a plant in field margin sharing pollinators with crops. The agronomic importance of this motif is enhanced pollination by shared pollinators and thereby increasing crop productivity. In turn, it also helps us in understanding the effective role played by natural or semi-natural vegetation surrounding an agroecosystem (Simmons et al., 2019).

2.3. The Stability of an Ecological Network

The community (Jacobian) matrix describes species abundance dynamics over continuous time estimated at equilibrium conditions and is useful in the study of local asymptomatic stability of the equilibrium (Haydon, 1994). The stability of an ecological network is indicated by eigenvalue with the largest real part of the matrix. The equilibrium becomes unstable if the real part of the leading eigenvalue is positive and consequently the system converges into another ecological regime and some species in the community may lead to extinction (Gross et al., 2009). By contrast, small perturbations around the equilibrium do not have any significant effect on the system and therefore the system returns to a stationary regime. Resilient systems tend to return to equilibrium rapidly after perturbation. This property of resilience is measured by the absolute value of the negative leading eigenvalue of the community matrix (Neubert & Caswell 1997). We can also measure the stability of a system by the removal of a specific species from a system. The consequent loss of additional species after the removal of a particular species is known as the extinction cascade. A system resists extinction cascade using its robustness (Mougi, 2016). In addition, the stability of a system is measured by resistance to invasion by alien species. Network robustness can be studied at two different levels: node-level robustness and edge-level robustness. A node-level robustness study requires perturbation of vertices whereas edges between nodes are perturbed to study edge-level robustness. The robustness of the network is enhanced by alterations in the topology of an ecological network (Dunne et al., 2002b). In addition, the formation of new interactions between species also leads to enhanced robustness in a community.

2.4. Databases with Useful Resources on Ecological Networks

The Interaction Web Database (IWDB) (http://www.ecologia.ib.usp.br/iwdb/) is a collaborative effort of researchers interested in species interactions in different ecosystems. Although this database was created by researchers in the United States in 2003, it is currently maintained by a group of researchers in Brazil. The Web of Life (https://www.web-of-life.es/) is another important database providing access to a dataset of ecological interactions (Fortuna et al., 2014). This database is maintained by the Bascompte laboratory at the University of Zurich. It provides small circles of different colors representing different types of ecological networks over the world map. For example, plant-herbivore interactions are shown in green color whereas yellow color indicates pollination. A mouse click on the circle opens a datasheet window containing full details regarding species interactions such as network ID, number of species, number of interactions, matrix and name of the species present in the network. Global Biotic Interactions (GloBI) is an open-access resource to find species interaction data such as prey-predator, plant-pollinator and so on. (Poelen et al., 2014). It is developed by a group of bioinformaticians, computer scientists and biologists. The R community can access the data using the R package rglobi (https://cran.r-project.org/web/packages/rglobi/index.html). The Global web database (https://www.globalwebdb.com/) provides access to only published food webs having direct trophic relationships across different taxa. This database is hosted by the University of Canberra and has more than 360 food webs in Excel formats (Thompson et al., 2012).

3. R Language and Environment

R is an open-source programing language and environment for statistical computing (Tiwary, 2022). The Comprehensive R Archive Network is a comprehensive project to store R binaries and packages (Hornik, 2012). An R package can be installed in an R environment by choosing a specific repository on the Comprehensive R Archive Network Web site. The loading of a package in the R environment is performed by the command, *library (package_name)*. There are more than 18,000 R packages available for statistical computing to date. R scripts are text files having R commands to perform a specific task. There are many R packages available for graphics as well.

3.1. R packages Useful for Analysis of Ecological Networks

There are some useful R packages for the analysis of ecological networks. The igraph (https://igraph.org/) is an open-source R package for network analysis. It can handle a large number of nodes and edges with powerful visualization using plot and tkplot functions. Two packages namely rmangal and rglobi are widely used to access mangal and globi databases respectively in the R environment (Poisot et al., 2016). The rmangal package automates the analysis process and improves replicability and reproducibility of ecological network studies (Por & Dor, 1984). The rglobi can search species interactions by location, interaction type, and taxonomic name. The vegan package (https://cran.r-project.org/web/packages/vegan/index.html) provides various multivariate tools for descriptive community analysis such as diversity and dissimilarity analysis (Dixon, 2003). Another R package for multivariate data analysis is ade4 having a useful large number of functions including principal-component analysis and correspondence analysis (Dray & Dufour, 2007; Hornik, 2012). The graphical user interface of ade4 is known as ade4TkGUI with essential methods available in ade4 (Thioulouse et al., 2018). The hmsc package uses a technique to analyze community data in ecology. Hierarchical modeling of species communities provides a general and flexible framework for fitting joint species distribution modeling. This package can integrate the community ecology data with metadata information such as environmental covariates, species traits, phylogenetic relationships and the spatiotemporal context of the study. Overall, it provides community assembly processes from non-manipulative observational data of species communities by predictive modeling (Tikhonov et al., 2020). A package known as econullnetr implements null modelling and plotting functions for ecological networks. It also provides data-export tools which are compatible with existing network analysis packages (Vaughan et al., 2018). The bmotif is an R package designed for motif analyses in bipartite networks designed on a mathematical framework to derive formal expressions for motif frequencies (Simmons et al., 2019). The frequencies of species occurrence in various positions within motifs can be also measured. Three R packages ggbipart, Cheddar and foodweb are extensively used for the visualization of ecological networks.

3.2. Analysis of Ecological Network Using R Package Igraph

First, we will create an adjacency matrix of a food web. The position of each element corresponds to a specific organism in the food web. For example, Row 1 and Column 1 both refer to plants, a producer in the food web. The element a_{21} = 1 means that the insect (organism 2) feeds on the plant (Organism 1). Tiger is the top predator in this food web because it feeds on most animals. The matrix of foodweb in the form of a Table is as follows:

This matrix can be saved as comma separated value (CSV) file with the name “foodweb.csv” and included in the current working directory of R. The above-mentioned dataset can be imported from the “foodweb.csv” file in R using the following command:

	Plant	Insect	Bird	Fish	Rat	Deer	Tiger
Plant	0	1	1	1	1	1	0
Insect	0	0	1	1	1	0	0
Bird	0	0	0	0	0	0	1
Fish	0	0	1	0	0	0	1
Rat	0	0	0	0	0	0	1
Deer	0	0	0	0	0	0	1
Tiger	0	0	0	0	0	0	0

>foodweb<- read.csv(‘foodweb.csv’, sep= ‘,’ , header = T, row.name=1)

The above R command creates an object “foodweb” with function *read.csv* along with three arguments regarding the nature of the separator (comma), header (TRUE), and the presence of row name in the first column. The data is first read as a data frame in R which is subsequently converted to a matrix using the following command:

>foodweb<-as.matrix(foodweb)

Now, we will install the igraph package in the R environment using the following command:

install.packages(“igraph”)

The package is loaded into the workspace using the command which is as follows:

library(igraph)

The adjacency matrix is converted to the igraph network using the following R code.

foodweb.network<-graph.adjacency(foodweb)

All species are represented in both rows and columns and the number of species (species richness) of an adjacency matrix is computed as.

```
> S<-dim(foodweb)[1]
> S
[1] 7
```

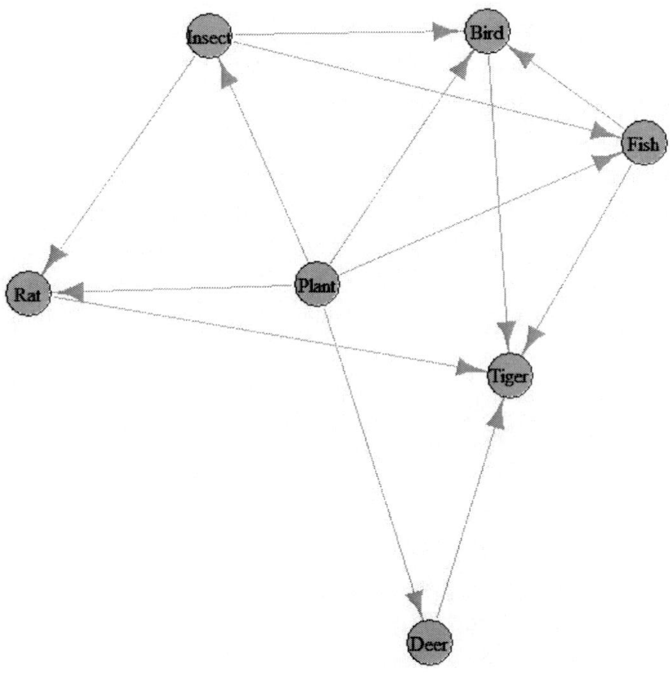

Figure 5. A food web network showing the connection between different species.

Since the interaction in each element is labeled as one in the matrix, the total number of interactions can be obtained with a sum of all the values in the matrix using the following R code.

```
> L<-sum(foodweb)
> L
[1] 13
```

First, the adjacency matrix is converted to an igraph network and then the network is drawn using the command *plot*

```
> foodweb. network<-graph.adjacency(foodweb)
> plot(foodweb.network)
```

Figure 5

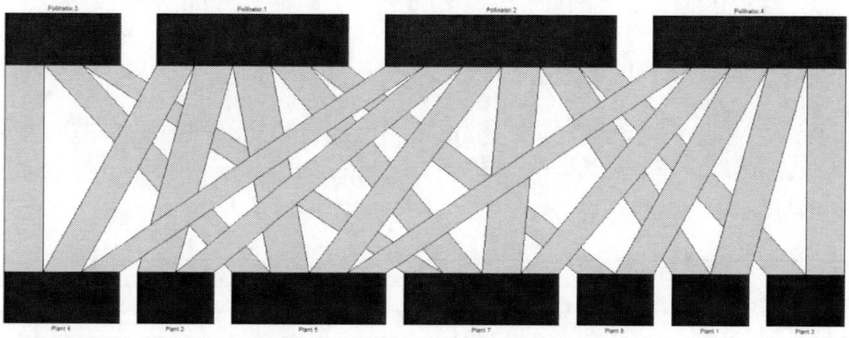

Figure 6. A bipartite graph plotted using the function plotweb.

Now, again we count the total number of species (S) and the total number of interactions (L) using the following code.

```
S<-vcount(foodweb.network)
> S
[1] 7
> L<-ecount(foodweb.network)
> L
[1] 13
```

The connectivity (L/S) and connectance (L/S^2) of the food web are calculated as

```
>Connectivity<-L/S
> Connectivity
[1] 1.857143
> C<-L/S^2
> C
[1] 0.2653061
```

3.3. Analysis of Bipartite Network Using Two R Packages: Igraph and Bipartite

The counterpart of the adjacency matrix in the bipartite network is known as the incidence matrix. Here, one set of species such as plants is represented as rows whereas another set of species such as pollinators is represented as columns. The size of the incidence matrix is represented as H X V (No. of hosts X No. of Visitors). For example, plant-pollinator interactions can be represented in the following table. The interaction between a plant and a pollinator is indicated by one in the incidence matrix.

	Pollinator 1	Pollinator 2	Pollinator 3	Pollinator 4
Plant 1	0	1	0	1
Plant 2	1	1	0	0
Plant 3	0	1	0	1
Plant 4	1	1	1	0
Plant 5	1	1	1	1
Plant 6	0	0	0	0
Plant 7	1	1	1	1
Plant 8	1	0	0	1

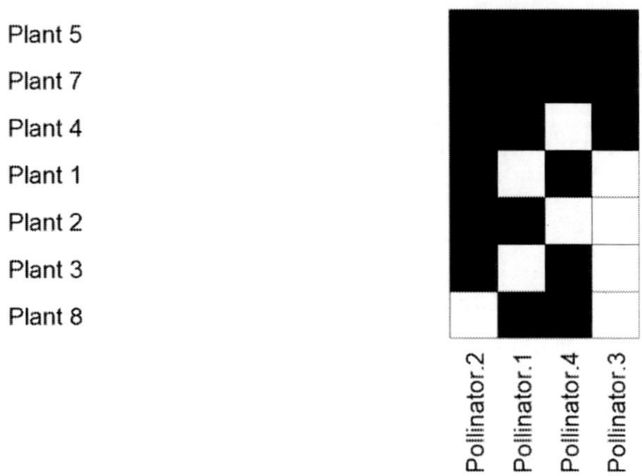

Figure 7. An adjacency matrix plotted using the function visweb. The dark cell indicates interaction between pollinator and plant.

The number of pollinators in the network is the number of rows in the matrix.

```
> pollination<- read.csv(‘pollination.csv’, sep= ‘,’ , header = T, row.name=1)
> pollination<-as.matrix(pollination)
> Plants<-dim(pollination) [1]
> Plants
[1] 8
> Pollinators<-dim(pollination) [2]
> Pollinators
[1] 4
```

The total number of species involved in interactions is as follows.

```
> S<-Pollinators+Plants
> S
[1] 12
```

The total number of plant-pollinator interactions is the sum of the incidence matrix.

```
> L<-sum (pollination)
> L
[1] 19
> library(bipartite)
> par(xpd=T)
> plotweb(pollination)
```

Figure 6

```
> visweb(pollination)
```

Figure 7

We can compute indices to characterize a network at various hierarchical levels such as network level, group level, link level, node level and species level. The indices computed for the network level are computed for the entire network and important network indices are explained in Table 3. The data frame returned after the *network-level* function is as follows:

```
network-level(pollination)
        connectance
        6.785714e-01
        cluster coefficient
        7.142857e-01
        weighted NODF
        0.000000e+00
        Fisher alpha
```

4.294967e+09
number.of.species.HL
4.000000e+00
cluster.coefficient.LL
7.500000e-01
togetherness.HL
1.527778e-01
V.ratio.LL
1.000000e+00
robustness.HL
5.421429e-01
partner.diversity.LL
1.049019e+00

web asymmetry
-2.727273e-01
modularity Q
1.855956e-01
interaction strength asymmetry
0.000000e+00
Shannon diversity
2.944439e+00
number.of.species.LL
7.000000e+00
weighted.cluster.coefficient.HL
6.324786e-01
togetherness.LL
1.309524e-01
discrepancy.HL
4.000000e+00
robustness.LL
7.339340e-01
generality.HL
5.000000e+00
links per species
1.727273e+00
nestedness
2.935118e+01
specialization asymmetry
-3.436790e-02
interaction evenness
8.836309e-01

mean.number.of.shared.partners.HL
3.166667e+00
weighted.cluster.coefficient.LL
9.569892e-01
C.score.HL
5.625000e-01
discrepancy.LL
3.000000e+00
functional.complementarity.HL
6.967632e+00

vulnerability.LL
3.000000e+00
number of compartments
1.000000e+00
NODF
6.209877e+01
linkage density
4.000000e+00
Alatalo interaction evenness
1.000000e+00
mean.number.of.shared.partners.LL
1.809524e+00
niche.overlap.HL
6.618687e-01
C.score.LL
4.250000e-01
extinction.slope.HL
3.852553e+00
functional.complementarity.LL
6.608666e+00

compartment diversity
NA
weighted nestedness
2.057735e-01
weighted connectance
3.636364e-01
H2
0.000000e+00
cluster.coefficient.HL
7.142857e-01

```
niche.overlap.LL
6.451247e-01
V.ratio.HL
1.000000e+00
extinction.slope.LL
7.358753e+00
partner.diversity.HL
1.586356e+00
```

Table 1. Useful online open-access resources providing data for ecological networks

Database	Description	URL
Interaction web database	Species interactions from different parts of the world	http://www.ecologia.ib.usp.br/iwdb/
Web of Life	A dataset of different types of ecological interactions	https://www.web-of-life.es/
The Global Biotic Interactions Database (GloBI)	A database providing species interaction data	https://www.globalbioticinteractions.org/
The Global Web database	A database providing online access to food webs across different biomes	https://www.globalwebdb.com/
Mangal	Ecological Interactions database	https://mangal.io/

Table 2. R packages for ecological network analysis

R Package	Description
igraph	A widely used package for network analysis
rmangal	An R package for access to the Mangal database.
rglobi	An R package to access the GloBI database
vegan	Analysis of community matrix
ade4	Multivariate analysis of ecological data
hmsc	A model-based approach to analyse ecological community
econullnetr	Tools to analyse null models in ecological networks
bipartite	Visualization and analysis of bipartite web ecological networks
bmotif	Motif analysis (counting and species positions in a motif) of a bipartite network
ggbipart	Plotting functions for bipartite network
Cheddar	Analysis and visualization of ecological community
foodweb	Creates 3D graphs of foodwebs

Table 3. Common network level indices of ecological networks

Index	Description
Connectance	The proportion of realized interactions out of all possible interactions by all species
Web asymmetry	Balance of the number of species in the two levels
Cluster coefficient	It is a measure of clustering tendency of nodes in a network

Index	Description
Nestedness	It indicates the temperature of the matrix with zero indicating perfect nestedness.
Weighted nestedness	A weighted interaction nestedness estimator with one indicating perfect nestedness.
Weighted NODF	An alternative measure of quantitative nestedness with one indication perfect nestedness.
Interaction strength asymmetry	A measure of interaction between specialized species and generalized species
Specialization asymmetry	A measure of asymmetry in terms of a specialized species interacting with a generalist species.
Linkage density	Marginal-totals weighted diversity of interactions per species
Fisher alpha	A diversity index to compare communities with different numbers of individuals and it is independent of sample size.
Interaction evenness	A measure of balance in the distribution of interactions between species.
Shannon diversity	It estimates species diversity considering both the richness and evenness of species in a habitat.
H_2	Network-wide frequency-based specialization index ranging from zero (lack of specialization) to one (fully specialized).

Conclusion

The complexity of an ecological network can be analyzed using available statistical computing resources in the R language and environment. This computational analysis will further pave the way for maintaining ecological stability as well as the conservation of keystone species in an ecosystem. The findings of the computational analysis will be useful for conserving the entire biodiversity based on systems-level interactions across the species present in a particular habitat.

References

Almeida-Neto, M., Guimarães, P., Guimarães, P. R., Loyola, R. D., and Ulrich, W. (2008). A consistent metric for nestedness analysis in ecological systems: Reconciling concept and measurement. *Oikos*, 117(8), 1227–1239. https://doi.org/10.1111/j.0030-1299.2008.16644.x. http://paperpile.com/b/oL7glu/fWUj.

Bascompte, J., Jordano, P., Melián, C. J., and Olesen, J. M. (2003). The nested assembly of plant-animal mutualistic networks. *Proceedings of the National Academy of Sciences of the United States of America*, 100(16), 9383–9387. https://doi.org/10.1073/pnas.1633576100.

Cagua, E. F., Wootton, K. L., and Stouffer, D. B. (2019). Keystoneness, centrality, and the structural controllability of ecological networks. *Journal of Ecology*, 107(4),

1779–1790. https://doi.org/10.1111/1365-2745.13147. http://paperpile.com/b/oL7glu/wTWn.

Camacho, J., Guimerà, R., and Nunes Amaral, L. A. (2002). Robust patterns in food web structure. *Physical Review Letters*, 88(22), 228102. https://doi.org/10.1103/PhysRevLett.88.228102. http://paperpile.com/b/oL7glu/pcJZ.

Chesson, P., and Kuang, J. J. (2008). The interaction between predation and competition. *Nature*, 456(7219), 235–238. https://doi.org/10.1038/nature07248. http://paperpile.com/b/oL7glu/WtYz.

Delmas, E., Besson, M., Brice, M. H., Burkle, L. A., Dalla Riva, G. V., Fortin, M. J., Gravel, D., Guimarães, P. R., Jr, Hembry, D. H., Newman, E. A., Olesen, J. M., Pires, M. M., Yeakel, J. D., and Poisot, T. (2019). Analysing ecological networks of species interactions. *Biological reviews of the Cambridge Philosophical Society*, 94(1), 16–36. https://doi.org/10.1111/brv.12433.

Dixon, P. (2003). VEGAN, a package of R functions for community ecology. *Journal of Vegetation Science*, 14(6), 927–930. https://doi.org/10.1111/j.1654-1103.2003.tb02228.x. http://paperpile.com/b/oL7glu/dOP9.

Dormann, C. F., Frund, J., Bluthgen, N., and Gruber, B. (2009). Indices, graphs and null models: Analyzing bipartite ecological networks. *Open Ecology Journal*, 2(1), 7–24. https://doi.org/10.2174/1874213000902010007.
http://paperpile.com/b/oL7glu/OLXt.

Dray, S., and Dufour, A. B. (2007). The Ade4 package: Implementing the duality diagram for ecologists. *Journal of Statistical Software*, 22(4), 1–20. https://doi.org/10.18637/jss.v022.i04.

Dunne, J. A., Williams, R. J., and Martinez, N. D. (2002a). Food-web structure and network theory: The role of connectance and size. *Proceedings of the National Academy of Sciences of the United States of America*, 99(20), 12917–12922. https://doi.org/10.1073/pnas.192407699. http://paperpile.com/b/oL7glu/NtM2.

Dunne, J. A., Williams, R. J., and Martinez, N. D. (2002b). Network structure and biodiversity loss in food webs: Robustness increases with connectance. *Ecology Letters*, 5(4), 558–567. https://doi.org/10.1046/j.1461-0248.2002.00354.x. http://paperpile.com/b/oL7glu/V5th.

Estrada, E. (2012). *The structure of complex networks: Theory and applications*. Oxford University Press.

Feng, W., and Takemoto, K. (2014). Heterogeneity in ecological mutualistic networks dominantly determines community stability. *Scientific Reports*, 4(August), 5912. https://doi.org/10.1038/srep05912.

Fortuna, M. A., Ortega, R., and Bascompte, J. (2014). *The Web of Life. ArXiv*. https://doi.org/10.48550/arXiv.1403.2575.

Gross, T., Rudolf, L., Levin, S. A., and Dieckmann, U. (2009). Generalized models reveal stabilizing factors in food webs. *Science*, 325(5941), 747–750. https://www.science.org/doi/10.1126/science.1173536.

Havens, K. (1992). Scale and structure in natural food webs. *Science*, 257(5073), 1107–1109. https://www.science.org/doi/10.1126/science.257.5073.1107. http://paperpile.com/b/oL7glu/xM0c.

Haydon, D. (1994). Pivotal assumptions determining the relationship between stability and complexity: An analytical synthesis of the stability-complexity debate. *American Naturalist*, 144(1), 14–29. https://doi.org/10.1086/285658. http://paperpile.com/b/oL7glu/vBYR.

Hooper, D. U., Chapin, F. S., Ewel, J. J., Hector, A., Inchausti, P., Lavorel, S., Lawton, J. H., Lodge, D. M., Loreau, M., Naeem, S., Schmid, B., Setälä, H., Symstad, A. J., Vandermeer, J., and Wardle, D. A. (2005). Effects of biodiversity on ecosystem functioning: A consensus of current knowledge. *Ecological Monographs*, 75(1), 3–35. https://doi.org/10.1890/04-0922.

Hornik, K. (2012). The comprehensive R archive network. *WIREs Computational Statistics*, 4(4), 394–398. https://doi.org/10.1002/wics.1212. http://paperpile.com/b/oL7glu/atAa.

Jordano, P., Bascompte, J., and Olesen, J. M. (2002). Invariant properties in coevolutionary networks of plant-animal interactions. *Ecology Letters*, 6(1), 69–81. https://doi.org/10.1046/j.1461-0248.2003.00403.x. http://paperpile.com/b/oL7glu/7WNs.

Kokkoris, G. D., Jansen, V. A. A., Loreau, M., and Troumbis, A. Y. (2002). Variability in interaction strength and implications for biodiversity. *Journal of Animal Ecology*, 71(2), 362–371. https://doi.org/10.1046/j.1365-2656.2002.00604.x. http://paperpile.com/b/oL7glu/TY5k.

MacArthur, R. (1955). Fluctuations of animal populations and a measure of community stability. *Ecology*, 36(3). https://doi.org/10.2307/1929601. http://paperpile.com/b/oL7glu/z9Sn.

May, R. M. (1972). Will a large complex system be stable? *Nature*, 238(5364), 413–414. https://doi.org/10.1038/238413a0. http://paperpile.com/b/oL7glu/Dm4Q.

May, R. M. (2019). *Stability and complexity in model ecosystems*. Princeton University Press.

Milo, R., Shen-Orr, S., Itzkovitz, S., Kashtan, N., Chklovskii, D., and Alon, U. (2002). Network motifs: Simple building blocks of complex networks. *Science*, 298(5594), 824–827. https://www.science.org/doi/10.1126/science.298.5594.824.

Montoya, J. M., Pimm, S. L., and Solé, R. V. (2006). Ecological networks and their fragility. *Nature*, 442(7100), 259–264. https://doi.org/10.1038/nature04927.

Mougi, A. (2016). The roles of amensalistic and commensalistic interactions in large ecological network stability. *Scientific Reports*, 6(July), 29929. https://doi.org/10.1038/srep29929.

Mougi, A., and Kondoh, M. (2012). Diversity of interaction types and ecological community stability. *Science*, 337(6092), 349–351. https://www.science.org/doi/10.1126/science.1220529. http://paperpile.com/b/oL7glu/EPge.

Ndour, M., Abbaoui, K., Ammar, H., and Cherruault, Y. (1996). An example of an interaction model between two species. *Kybernetes*, 25(4), 106–118. https://doi.org/10.1108/03684929610118381. http://paperpile.com/b/oL7glu/2a61.

Neubert, M. G., and Caswell, H. (1997). Alternatives to resilience for measuring the responses of ecological systems to perturbations. *Ecology*, 78(3), 653–665. https://doi.org/10.1890/0012-9658(1997)078[0653:ATRFMT]2.0.CO;2. http://paperpile.com/b/oL7glu/knYO.

Newman, M. (2010). *Networks: An introduction.* Oxford University Press.

Novak, M., Yeakel, J. D., Noble, A. E., Doak, D. F., Emmerson, M., Estes, J. A., Jacob, U., Tinker, M. T., and Wootton, J. T. (2016). Characterizing species interactions to understand press perturbations: What is the community matrix? *Annual Review of Ecology, Evolution, and Systematics*, 47(1), 409–432. https://doi.org/10.1146/annurev-ecolsys-032416-010215. http://paperpile.com/b/oL7glu/vR4g.

Okuyama, T., and Holland, J. N. (2008). Network structural properties mediate the stability of mutualistic communities. *Ecology Letters*, 11(3), 208–216. https://doi.org/10.1111/j.1461-0248.2007.01137.x.

Olesen, J. M., Bascompte, J., Dupont, Y. L., and Jordano, P. (2007). The modularity of pollination networks. *Proceedings of the National Academy of Sciences of the United States of America*, *104*(50), 19891–19896. https://doi.org/10.1073/pnas.0706375104. http://paperpile.com/b/oL7glu/FNbl.

Olito, C., and Fox, J. W. (2015). Species traits and abundances predict metrics of plant-pollinator network structure, but not pairwise interactions. *Oikos*, 124(4), 428–436. https://doi.org/10.1111/oik.01439. http://paperpile.com/b/oL7glu/Qb5B.

Pascual, M., Dunne, J. A., and Dunne, J. A. (2006). *Ecological networks: Linking structure to dynamics in food webs*. Oxford University Press.

Pimm, S. L., Lawton, J. H., and Cohen, J. E. (1991). Food web patterns and their consequences. *Nature*, 350(6320), 669–674. https://doi.org/10.1038/350669a0. http://paperpile.com/b/oL7glu/72ZD.

Poelen, J. H., Simons, J. D., and Mungall, C. J. (2014). Global biotic interactions: An open infrastructure to share and analyze species-interaction datasets. *Ecological Informatics*, 24, 148–159. https://doi.org/10.1016/j.ecoinf.2014.08.005. http://paperpile.com/b/oL7glu/1Uwt.

Poisot, T., Baiser, B., Dunne, J. A., Kéfi, S., Massol, F., Mouquet, N., Romanuk, T. N., Stouffer, D. B., Wood, S. A., and Gravel, D. (2016). Mangal – Making ecological network analysis simple. *Ecography*, 39(4), 384–390. https://doi.org/10.1111/ecog.00976. http://paperpile.com/b/oL7glu/VQ7K.

Por, F. D., and Dor, I. (1984). *Hydrobiology of the mangal: The ecosystem of the mangrove forests*. Springer.

Schoener, T. W. (1989). Food webs from the small to the large: The Robert H. MacArthur award lecture. *Ecology*, 70(6), 1559–1589. https://doi.org/10.2307/1938088. http://paperpile.com/b/oL7glu/POSw.

Simmons, B. I., Cirtwill, A. R., Baker, N. J., Wauchope, H. S., Dicks, L. V., Stouffer, D. B., and Sutherland, W. J. (2019). Motifs in bipartite ecological networks: Uncovering indirect interactions. *Oikos*, 128(2), 154–170. https://doi.org/10.1111/oik.05670. http://paperpile.com/b/oL7glu/rwKX.

Simmons, B. I., Sweering, M. J. M., Schillinger, M., Dicks, L. V., Sutherland, W. J., and Di Clemente, R. (2019). Bmotif: A package for motif analyses of bipartite networks. *Methods in Ecology and Evolution*, 10(5), 695–701. https://doi.org/10.1111/2041-210X.13149. http://paperpile.com/b/oL7glu/ibfu.

States, D. J. (2004). Faculty opinions recommendation of finding and evaluating community structure in networks. *Faculty opinions – Post-publication peer review of the biomedical literature*.

Thioulouse, J., Dray, S., Dufour, A. B., Siberchicot, A., Jombart, T., and Pavoine, S. (2018). *Multivariate analysis of ecological data with ade4*. Springer.

Thompson, R. M., Brose, U., Dunne, J. A., Hall, R. O., Jr. Hladyz, S., Kitching, R. L., Martinez, N. D., Rantala, H., Romanuk, T. N., Stouffer, D. B., and Tylianakis, J. M. (2012). Food webs: Reconciling the structure and function of biodiversity. *Trends in Ecology and Evolution*, 27(12), 689–697. https://doi.org/10.1016/j.tree.2012.08.005.

Tikhonov, G., Opedal, Ø. H., Abrego, N., Lehikoinen, A., de Jonge, M. M. J., Oksanen, J., and Ovaskainen, O. (2020). Joint species distribution modelling with the R-package Hmsc. Methods in Ecology and Evolution. *British Ecological Society*, 11(3), 442–447. https://doi.org/10.1111/2041-210X.13345.

Tiwary B. K. (2022). *Bioinformatics and Computational Biology*. Springer; Singapore. Statistical Computing Using R.; pp. 33–52. https://doi.org/10.1007/978-981-16-4241-8_3.

van Altena, C., Hemerik, L., and de Ruiter, P. C. (2016). Food web stability and weighted connectance: The complexity-stability debate revisited. *Theoretical Ecology*, 9(1), 49–58. https://doi.org/10.1007/s12080-015-0291-7. http://paperpile.com/b/oL7glu/LVoG.

Vaughan, I. P., Gotelli, N. J., Memmott, J., Pearson, C. E., Woodward, G., and Symondson, W. O. C. (2018). Econullnetr: An R package using null models to analyse the structure of ecological networks and identify resource selection. *Methods in Ecology and Evolution*, 9(3), 728–733. https://doi.org/10.1111/2041-210X.12907.

Chapter 4

Forests: A Valuable Natural Resource: Its Degradation and Restoration Perspectives

Sunil Singh[1,*]
and Rajani Srivastava[2]
[1]Department of Botany, Udai Pratap College, Varanasi, Uttar Pradesh, India
(Mahatma Gandhi Kashi Vidhyapeeth, Varanasi)
[2]Department of Environment and Sustainable Development, IESD,
Rajiv Gandhi South Campus, Banaras Hindu University, Mirzapur, India

Abstract

In general, forests are self-sustained wooden tracts dominated by trees forming a sort of canopy. Forests are a valuable resource both economically and ecologically, besides having aesthetic value. However, widespread deforestation of the world's forests has resulted in environmentally and aesthetically poor landscapes around the globe. This has resulted in forest degradation. A degraded forest is one where the forest cover has been affected by unsustainable exploitation of timber and non-timber products to such an extent that its capacity to recover completely from such alteration is compromised. The degradation of a forest may be evaluated based on three key elements, i.e., (1) a reference forest, (2) degradation indicators, and (3) a threshold value to determine the point at which a non-degraded forest becomes a degraded forest. FAO's Forest Resources Assessment (2018) estimated that the world's forest area decreased by about 3% between 1990 and 2015. This has resulted in widespread forest restoration and regeneration activities all over the world. However, a successful restoration faces many challenges, like high cost, a shortage of

* Corresponding Author's Email: upcvns@gmail.com.

In: Conserving, Halting and Reinstating Degraded Natural Resources …
Editors: Rajani Srivastava and Pankaj Kumar
ISBN: 979-8-89113-067-8

regeneration materials, and the usefulness of restoration for local and regional communities. However, a successful large-scale restoration should keep the following factors in mind: (1) learning from the past; (2) ecological knowledge supporting forest restoration; (3) advances in restoration and regeneration techniques and systems; (4) forest restoration following biotic and abiotic disturbances; and (5) restoring forest landscapes for the future. The concept of New Generation Plantations (NGP) is one promising way forward for successful restoration; however, such multipurpose approaches create new challenges for research and management.

Keywords: forest, degraded forest, deforestation, forest degradation, reference forest, new generation plantation

1. Introduction

Widespread deforestation of the world's forests has resulted in environmentally and aesthetically poor landscapes around the globe. The effects of deforestation, to a greater extent, have been offset through natural regeneration of forest and the establishment of plantations. However, the regenerated forests dominated by a few species and plantations are designed to yield one or two products, neglecting the goods and services that a natural forest can contribute to the well-being of local communities (Lamb & Gilmour, 2003).

In general, forests are self-sustained wooden tracts with a biotic community dominated by trees forming a sort of canopy. Food and Agricultural Organization (FAO) defines forest as "a land spanning more than 0.5 ha with trees higher than 5 m and a canopy cover of more than 10 percent, or trees able to reach these thresholds *in situ*" (FAO, 2013). The word "forest" is derived from Latin "*foris,*" meaning "outside the doors." Roughly, they contain 90% of the terrestrial biomass and cover about one third of land. In total, they cover 4.06 billion ha (31 percent of the world's land surface), but this area is decreasing, particularly in the tropics (FAO, 2022). The tropical part of the earth has the largest proportion of the world's forests (45%), followed by the boreal, temperate, and subtropical parts. More than half (54%) of the world's forests are in only five countries: the Russian Federation, Brazil, Canada, the United States, and China (FAO, 2020). In India, forest cover covers approximately 775288 km^2 (23.88% of the land) (State of Forest Report, 2021). Forests are valuable both economically and ecologically, besides having aesthetic value. Economically, forests are a source of wood, food articles, spices, essential oils, camphor, tannins, gums, resins, dyes, cork, and valuable drugs. Ecologically, forests provide plant cover for the soil, maintain subsoil water, moderate the climate by increasing the frequency of rainfall, provide a home for plants and animals, and reduce atmospheric pollution too. Although forests cover 31 percent of the Earth's land surface (4.06 billion ha), the area is shrinking, with 420 million hectares of forest lost through deforestation between 1990 and 2020. The rate of deforestation is declining, but it was still 10 million ha per year in 2015–20. Some 47 million ha of primary forests were lost between 2000 and 2020 (FAO, 2022).

1.1. Forest Degradation and a Degraded Forest

Forest degradation is an ongoing process that has resulted in a degraded forest. The Food and Agriculture Organization of the United Nations (FAO) has defined "forest degradation" as changes within a forest that affect the structure and function of the forest stand or site and thereby lower its capacity to supply products or services. Thus, "forest degradation" refers to forests that have lost their former ecological integrity. The term "degraded forest" is applied to all forests where the forest cover has been affected due to unsustainable exploitation of timber and non-timber products to such an extent that its capacity to recover completely from such alteration is compromised. Degraded forests are distinguished by changes in the ecosystem's structure, process, function, and dynamics. Degraded forests provide a reduced output of products and services for a given area and a limited degree of biological diversity (IITO, 2002). In practice, the state of a stand is determined according to its dominant floristic composition and structure, including height, strata, and density. Thus, a stand that was once dominated by a group of species but has since changed due to human-induced disturbances (e.g., regeneration, growth rates, mortality, decomposition) becomes a different type of ecosystem than that expected at that site in any of its successional states (Thompson et al., 2009; Thompson et al., 2011). These changes constitute a degraded forest if they include at least several characteristics such as loss of the main canopy, the dominance of competitive species, loss of functional groups, the spread of generalist and invasive species, and a generally impoverished biota (Thompson et al., 2012), and a reduction in the source of propagules from the original vegetation. The latter is the most critical for degraded forests, as the capacity of these tree species to regenerate following a disturbance is reduced (Stanturf et al., 2014). A significant reduction in the population of a given tree species may trigger the loss of other associated species and a reduction in the provision of specific goods and services (Devi & Behera, 2003).

Forests may be partially or severely degraded, depending on the degree to which the structural and compositional characteristics of the forest have been modified and the frequency and intensity of the human-induced impacts. Partially degraded forests have only minor changes in structure and composition, but this is enough to severely reduce the regeneration capacity of key species due to a lack of seed sources or the failure of seedlings to develop due to grazing. On the other hand, severely degraded forests are characterised by certain competitive shrub and tree species (native or invasive) growing to such an extent that they hinder natural regeneration and result in a state of arrested succession. In the most extreme cases of degradation, changes may occur not only in the biotic environment but also in the physical environment, such as the soil (Chazdon, 2008; Ghazoul et al., 2015). In all cases, the forest regeneration process may be the key indicator of the degree of degradation. The size and diversity of seed banks, shoots, and seedlings are inversely proportional to the degree of alteration that the forest has experienced. In other words, as change becomes more serious or frequent, the reduction in abundance and diversity of sources of propagules also becomes greater. Often, these sources may be lost completely, creating a major barrier to the regeneration process (Holl et al., 2000).

All factors that are responsible for forest degradation may be classified into two categories: natural factors, like water and nutrient deficiency, diseases, aging and senescence, and deforestation for the expansion of agriculture or rangelands; and anthropogenic factors, like the exploitation of forests for forest products, the dispersion of pollutants, and overloaded with nutrients. Forest degradation has now been aggravated all over the world due to the direct and indirect effects of a variety of anthropogenic activities. The direct effects include, for example, the replacement of a forest area by a field of some cash-crop trees, while the indirect effects include the construction of roads in a forest zone that open the gate for the search for new cultivation land by people.

1.2. Defining Degradation

Forest degradation results in biodiversity loss, greenhouse gas emissions, and the diminution of ecosystem goods and services, but scientists, practitioners, and policymakers are unable to agree on a framework for defining degradation. However, for the evaluation of forest degradation, three key elements may be considered: (1) a reference forest; (2) degradation indicators; and (3) a threshold value to determine the point at which a non-degraded forest becomes a degraded forest.

1.2.1. Forests of Reference

In assessing forest degradation, a reference or baseline can be used. It is done by comparing the present state of the forest either with a previous state of the forest before the occurrence of any impact or with a point of reference that must represent non-degraded forests with biophysical characteristics that are comparable to the forest being evaluated (that is, a local reference point) (Morales-Barquero et al., 2014). In practice, unmanaged primary or old-growth forests may provide a baseline for measuring degradation, even at different stages of succession (FAO, 2011; Lindenmayer et al., 2010; Thompson et al., 2009), and a sustainably managed natural forest may also provide a useful reference.

1.2.2. Degradation Indicators

These are the indicators that indicate the changes in the structure, composition, and regeneration of a forest. Multiple indicators are monitored for the evaluation of degradation (Lund, 2009; Morales-Barquero et al., 2014; Thompson et al., 2013). FAO (2011) proposes guidelines for degradation assessments that focus on the measurement of biophysical and biological effects and the direct causes of the degradation. FAO also advocates the use of certain key indicators, like stand volume, biomass stock, state and diversity of the ecosystem, functional and invasive species, and lack of or insufficient density of individuals, juveniles, and seedlings from commercial species. Thompson et al., (2013) propose a set of indicators to quantify types of forest degradation according to specific management situations and objectives, including growing stock, ecosystem state, tree species composition, invasive species, tree density, and the relative abundance of high wood density tree species. Stanturf et al., (2014) propose grouping indicators according to the forest attributes that have been affected by degradation. These include loss of canopy cover, reduction in structural complexity, reduction in growth, increase in mortality, a decline in biodiversity (e.g., loss of species of interest, the spread of invasive species), diminished protection functions (e.g., soil loss), and diminished production functions (e.g., a decline of commercial species, lack of regeneration). Modica et al., (2015) proposed the integrated forest degradation index, which is based on six indicators: Structural index (which measures tree stratification, i.e., their vertical distribution), canopy cover, natural regeneration density, focal species of degradation (i.e., typically associated species; functional species whose loss results in progressive changes to the forest, affecting many species in turn), coarse woody debris (i.e., volume per hectare), and soil depth are all factors to consider. By contrast with the above methodological approach, Ghazoul et al., (2015) propose the use of indicators linked to functional forest processes, including variables such as shoot growth in woody plants, seed availability in soil seed banks, an abundance of seedlings, and the age structure of common species (compared to that expected based on the successional state of the forest and the regeneration strategies of the various species). These indicators provide information regarding forest recovery in response to disturbances and, through analysis of the seedling and young tree communities, offer predictions as to the future dynamics of these forests.

1.2.3. Degradation Threshold

It is a value that marks the boundary between a non-degraded forest and a degraded forest and may refer to structural variables or indicators of the forest. Pragmatic criteria should be used to determine this threshold based on an understanding of the properties, patterns, and processes of forest ecosystems (Morales-Barquero et al., 2014). Tucker et al., (2008) state that quantitative measurements should be made of variables that describe the states of forests. While Baland et al., (2010) indicate that forest degradation has several dimensions that are difficult to represent in a single measure, however, they provide threshold values for several forest variables in the Indian Himalayas. Sasaki et al., (2011) propose a critical degradation threshold of 10% of forest cover in tropical forests.

2. Types of Degraded Forest Land

According to Lamb and Gilmour (2003), there are three broad categories of degraded forest lands.

First-category degraded forest land is the most severely degraded forest land that has lost much of its original biodiversity, structure, and biomass or site productivity (Point B in Figure 1).

Second-category degraded forest land is land occupied by woody regrowth that has developed on a site after an earlier disturbance, for example, agricultural clearing after which the land was abandoned. The site now has some of its former diversity, although it is dominated by early successional species and usually has fewer plant or animal species representative of mature forests (Point C in Figure 1).

The third category of degraded forest land includes degraded primary forests that have been severely disturbed, such as heavy logging, altering the forest structure and reducing biomass. However, some residual trees remain, and many primary forest tree species will be present as saplings or seedlings (see Figure 1).

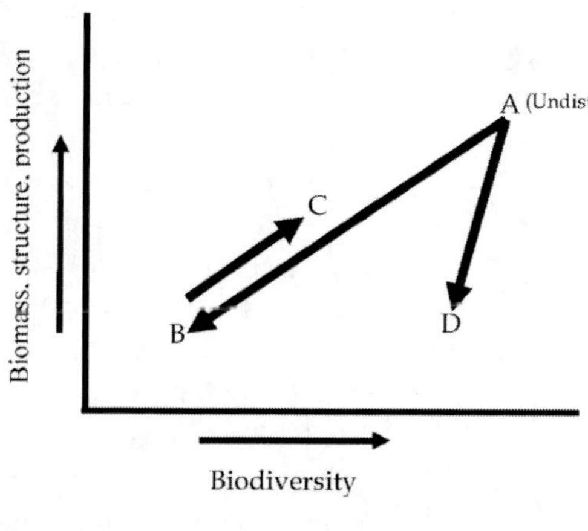

Figure 1. Different categories of forest degradation. An undisturbed forest is represented by Point A, where it has a certain level of biodiversity with biomass, structure and production. Extensive clearing can take it to Point B, where it will have lost a large proportion of its biodiversity as well as much of its biomass, structure and productivity. It may or may not recover unaided to Point C (e.g., becoming secondary regrowth forests). Some forms of intensive logging can cause the degradation of primary forests (A) while still leaving a significant amount of biodiversity (Point D) (Adapted from Lamb & Gilmour, 2003).

3. Need for Restoring Forests

Throughout the globe, forests provide a range of ecosystem services. International conventions and national policies for biodiversity conservation and climate change mitigation emphasize the need for increased forest protection, restoration, and management in response to climate change. However, global forest cover continues to decline. The Forest Resources Assessment (2018) estimated that the world's forest area decreased by about 3% between 1990 and 2015 (FAO, 2018). Against this background, forest restoration activities have become increasingly common around the world, and the regeneration of trees is a key component of these projects. The World Resource Institute in 2011 suggested that about 2 billion ha of land is suitable for different kinds of forest restoration (Minnemayer et al., 2011; Silva et al., 2019). Several global, regional, and national organizations have set targets during the last decade for large-scale forest landscape restoration. For example, The Bonn Challenge on forests, climate change, and biodiversity, launched by the Government of Germany and the IUCN in 2011, aims to restore 150 million hectares of degraded and deforested landscapes by 2020 and 350 million hectares by 2030. In 2014, the New York Declaration on Forests (NYDF) was adopted as a political declaration calling for the end of natural forest loss and the restoration of 350 million ha of degraded landscapes and forestlands by 2030 (Jacobs et al., 2015). Then, in 2021, at the UNFCCC COP26 in Glasgow, world leaders announced the Glasgow World Leaders Declaration on Forests and Land Use, a commitment aligned with the NYDF to end global deforestation by 2030, endorsed by over 140 countries and covering roughly 90% of global forests. The United Nations has recognized the critical role of ecosystem restoration as a tool for improving environmental conditions and enhancing human communities by designating 2021–2030 the United Nations Decade of Ecosystem Restoration (UN, 2019). Restoration ambitions are high, and these initiatives have received much support from countries and organizations around the world.

4. Forest Restoration Challenges

A successful large-scale restoration faces many challenges:

First, the costs of achieving restoration are often high. A conservative estimate of 2000 USD per hectare means that at least 700 billion USD is required to restore 350 million ha by 2030. This represents a significant barrier to scaling up restoration worldwide, and there is a risk that the interest from policymakers will wane over time as the costs and difficulties become apparent (Löf et al., 2019). Restoration costs vary depending on the methods used, from lower-cost alternatives using natural regeneration with native tree species to higher-cost approaches for active restoration using site preparation and planting. Recently, there has been a debate regarding the advantages, disadvantages, and degree of success of these approaches (e.g., Crouzeilles et al., 2017; Meli et al., 2017; Reid et al., 2018), but no one-size-fits-all solution regarding restoration methods exists. To achieve the ambitious goals, a diverse set of methods tailored to the local context must be used. Normally, costs for restoration increase with the degree of degradation of an ecosystem (Chazdon, 2008; Stanturf et al., 2001). Thus, if appropriate seed sources are available, natural regeneration with native tree species may represent a cost-effective option that may be applied across large areas. In other cases, when no such seed sources exist or when sites are too harsh, active restoration using planting (sometimes with non-natives) is the only option to establish forests (Jacobs et al., 2015).

Second, an important obstacle for successful active restoration is often a shortage of regeneration material (Cernansky, 2018). Restoration projects involve the re-establishment of native tree species, but there is often little knowledge about how, where, and when to obtain genetic material (species and provenances) with desired properties for the sites under restoration and how to collect, store, and pretreat seeds, followed by cost-efficient cultivation of the seedlings in nurseries. In addition, we have little experience with establishing such regeneration systems; efficient regeneration systems currently exist for only a few tree species, mostly for use in industrial plantations, and these techniques are not always suited for restoration programs (Löf et al., 2012). In addition, climate change raises questions about where different tree species may thrive in the future. Assisted migration, that is, the human-assisted movement of plants to more climatically suitable habitats, maybe one way forward (Dumroese et al., 2016; Stanturf et al., 2014).

A third major constraint for successful cost-effective restoration in the long term is that the restored forests need to be a positive resource for the local and regional communities. Thus, the success of forest restoration is heavily dependent on people and their ability to use restored forests in a variety of ways. The mounting demands from a growing human population increase the already huge pressure on the world's natural resources. If society's needs are not considered, the risk is that the public interest in sustainably protecting forests will be insufficient. Therefore, multiple-use forestry, such as multipurpose mixed forest systems and agroforestry systems that provide food and other benefits for people, is important. Sometimes the use of multipurpose tree leaves can also be useful in increasing the fertility of the soil (Srivastava & Singh, 2013; Srivastava & Singh, 2014). In addition, increasing evidence suggests that integrating exotic commercial tree species with native ones can support biodiversity, the environment, and local economies. One such approach is the New Generation Plantation (NGP) Platform which was launched in 2007 (Silva et al., 2019). This concept aims to complement timber production with effective maintenance and enhancement of ecosystems and simultaneously contribute to socioeconomic development and climate change mitigation. Plantations, with a current share of about 7% of total global forests, can help reduce pressure on natural forests (FAO, 2010) as they contribute to about 33% of the global demand for industrial round wood, a demand that is predicted to increase (Silva et al., 2019). The NGP concept is one way to, at least partly, meet these growing demands.

5. Forest Restoration after Biotic and Abiotic Disturbances

Disturbances, like windstorms, forest fires, and pest or disease outbreaks, are an essential and natural part of forest ecosystems. These disturbances drive the ecological processes that lead to the structural complexity of the forest ecosystem (Löf et al., 2019). Although the economics of fiber production is the main driver of management in commercial forests, there are opportunities to use alternative forestry methods, such as retention trees, variable density thinning, and prescribed burning that mimic natural disturbances, to promote heterogeneous forest stand structure and biodiversity (Puettmann et al., 2016; Löf et al., 2019). Degradation and loss of forests due to anthropogenic disturbances, however, have resulted in a significant reduction of forest biodiversity and the need for the restoration of millions of hectares of degraded forests worldwide (Lindenmayer, 2019). In degraded environments, such as grazing lands, abandoned croplands, or areas following severe wildfires, structural elements and sources of microsite diversity are often limited (Cortina et al., 2011), adding to restoration challenges. Mining sites are particularly difficult to restore to functioning forest ecosystems because of the need for massive landscape-level reconstruction of soils, hydrology, and biota, as well as the associated forest fragmentation, erosion, and disposal of mine tailing wastes (Stanturf et al., 2014; Macdonald et al., 2015). Simultaneously, global change, including an exponential increase in the introduction of invasive species and a rapidly shifting climate, puts unprecedented pressure on forests (Dumroese et al., 2016). The number of threatened and endangered forest tree species continues to rise, and the practical restoration of such species is complex due to the need for effective integration of technology, ecology, and society (Jacobs et al., 2013). Thus, the genetic conservation of tree species has become an urgent global necessity as this provides the essential basis for the adaptation and resilience of tree species to environmental stress and change (Potter et al., 2017). While knowledge of historical species range distributions can help guide restoration targets, shifts in frost hardiness zones, drought severity indices, and the presence or absence of species competitors will drive species' regional adaptiveness in the future. We are still in the infancy of navigating these dynamic and critical issues that will have long-lasting consequences for the success of forest restoration programs.

6. Restoration of Forests

Forest deterioration can be overcome using a variety of strategies. In certain circumstances, the goal is to recreate the previous ecosystem and recapture the old biodiversity; in other circumstances, the aim is to merely utilize the land for some beneficial activity, like agriculture. The degree to which each method allows for the restoration of the original biodiversity varies. They are comparable, though, in that they all aim to create fruitful and reliable new land use.

6.1. Ecological Knowledge for Forest Restoration

Ecological theory concerning plant-plant interactions has evolved over the last two decades, with significant implications for cost-effective restoration (Löf et al, 2019). In particular, it has been stressed that negative (competition) and positive (facilitation) interactions co-exist to a greater degree than previously thought. Earlier, there was a strong focus on negative interactions, and still, the most common vegetation management techniques during regeneration and restoration are measures that reduce competition from non-desired vegetation. Examples of such measures include prescribed fire, herbicides, and mechanical site preparation. Concepts such as the stress-gradient hypothesis (He et al., 2013) and nurse plants (Gómez-Aparicio, 2009) have, however, the potential to guide restoration toward methods that improve seedling survival and growth while simultaneously keeping restoration costs low. The former concept states that facilitation relative to competition increases in importance with the degree of abiotic and biotic stress, and the latter concept states that one plant species can facilitate the growth and development of another species. In addition, managing plant–plant interactions for restoration will be particularly relevant in scenarios of increasing drought due to climate change, with the potential for an increasing role of facilitation to secure regeneration. However, these concepts have so far most frequently been applied to restoration in dryland ecosystems dominated by perennial herbs or shrubs (Gómez-Aparicio et al., 2004). Knowledge is still lacking for forest ecosystems where vegetation dynamics are more complex, i.e., temperate or tropical regions. To succeed, it is not only necessary to select good combinations of nurse- and target-tree species with complementary species traits but also to manage plant densities and community composition as plant cover and competition increase with time. Direct or indirect biotic interactions involving herbivores may also play a fundamental role in cost-effective forest restoration. Selective ungulate browsing influences plant–plant interactions to such an extent that browse-tolerant or less palatable tree species often are favored and, thus, challenge restoration efforts to establish more species-rich forests (Kuijper et al., 2010; Metslaid et al., 2013). Physical or chemical protection of regeneration by fences, repellents, or individual tree shelters is usually effective, but the costs of establishing, maintaining, and removing these measures are high. Low-cost strategies may include providing physical protection for the target trees by using dense natural regeneration, direct seeding of species that are more browse-tolerant, or using nurse trees and shrubs. The ungulate impact can also be reduced by reducing populations (through hunting), increasing the carrying capacity of the forest landscape (by providing more food for the deer), or combining the two approaches. In addition, fodder crops of grasses, herbs, or woody species may be established to release browsing pressure from more desirable species on restoration sites, and supplementary feeding might be an option if this food improves nutrition for the ungulates (Felton et al., 2016). However, it may be difficult to translate new research findings about low-cost browsing-tolerant regeneration alternatives to restoration practices.

6.2. Restoration and Regeneration Techniques and Systems

Forest restoration and regeneration practices must evolve in response to changing societal demands, human and natural disturbances, policies, markets, technologies, and climate (Wagner et al., 2018). Much of the past science in this area emphasized reforestation following timber harvest for industrial purposes, but an increasing focus on the restoration of harsh, degraded environments demands new seedling production and planting techniques (Oliet & Jacobs, 2012). An improved understanding of biotic and abiotic factors that affect the ecophysiology of plant establishment is required to overcome inherent challenges on restoration sites. Nursery systems aim to produce high-quality, stress-resistant planting stock that will withstand stresses following field establishment (Haase & Davis, 2017). Stock types, fertilization or irrigation regimes, and hardening practice can be manipulated in the nursery to tailor morphological and physiological seedling attributes to optimize performance under the conditions of a given planting site (Grossnickle & El-Kassaby, 2015; Dumroese et al., 2016). Seedling size is positively correlated with plant content of nitrogen and carbohydrates, and seedlings may mobilize these stored reserves to resist environmental stresses on the planting site (Villar-Salvador et al., 2015). Site preparation is often necessary on restoration sites to ensure natural or artificial regeneration success, with requirements often dependent upon the severity of the prior disturbance. For instance, severe landscape disturbance associated with mine reclamation may require novel innovations in geomorphic restoration and site reconstruction (MacDonald et al., 2015; Zapico et al., 2018) to ensure functioning soil. Technological advances in mechanical site preparation have improved the efficiency of controlling site limiting factors (i.e., target vegetation, compaction, mounding), costs, environmental impact, and worker safety (Löf et al., 2012; Löf et al., 2019). Continued enhancements in the application of field fertilization, soil amendments, and tree shelters (Rose et al., 2016; Oliet et al., 2019) will further promote regeneration and restoration success. While fencing is generally effective against ungulate damage, the high cost of installation and maintenance limits its practical use. Thus, continued evaluation of alternatives such as tree shelters, nurse plants, and silvicultural planting schemes that account for browse risk (Burney & Jacobs, 2018; Maltoni et al., 2019) will help to ensure cost-effective restoration.

According to Lamb & Gilmour (2003), the following methods can be applied to stop the degradation of forests:

1. *Reclamation.* By primarily using exotic tree species, it is regaining productivity at a damaged forest site. Monocultures of one species are common. Original biodiversity is not recovered, but protective functions and many of the original ecological services may be re-established.
2. *Rehabilitation.* It is restoring some, but not all, of the original plant and animal species, as well as their production. The new forest may contain species that weren't there at first for ecological or economic reasons. The ecological services and protective role of the original forest may be restored over time.

Environmental Restoration It is restoring the forest's original composition, productivity, and species variety. Ecological functions and processes will eventually resemble those of the original forest. It is "the process of helping the recovery of an ecosystem that has been degraded, damaged, or destroyed," according to the Society for Ecological Restoration.

7. Restoration Methods

The following methods can be opted for restoration, rehabilitation, and reclamation.

7.1. Passive Restoration by Disturbed Site Protection

The extremes of a restoration action continuum are passive restoration, which solely relies on natural processes, and technical reclamation, which heavily interferes with the restoration process. Smaller sites recovering from moderate disruptions with low abiotic stress and moderate productivity should often favor passive restoration (Aerts & Honney, 2011). Passive restoration can be achieved simply by protecting the site from further disturbances and allowing natural colonization and successional processes to restore ecosystem biodiversity and structure (Shono et al., 2007). This type of restoration is applied to the forest where low wood density has been found due to wood removal for fuel and grazing. To reduce fuel wood extraction, it is necessary to evolve alternative sources, such as the development of agro-forestry and community wood lots, in this vital area.

Passive restoration is advantageous when there are limited financial resources available. This approach has several potential disadvantages, like the fact that eradication of weeds and pests can be difficult and expensive, and failure to remove them can limit the natural regeneration of native species and successional development. If the original disturbance permanently altered the local environment (e.g., topsoil has been lost), full recovery may be impossible with this method.

7.2. Supplementing Biodiversity by Afforestation

All secondary woods don't have a high level of diversity since they've been disturbed so frequently in the past that only a few common species are left. In these situations, extensive reforestation involving many important species is necessary (Martins, 2018). Due to compaction and erosion, the soil is typically in poor condition in these places. The use of species that may increase soil structure and fertility by forming quick root growth and supporting high root turnover would speed the process of reforestation. For instance, soil fertility is improved by trees that fix nitrogen. Root nodules of *Acacıa be them, A. catectic, Albizia lebbeck, Leucaena leucocephala, Mimosa pudica*, and *Prosopis cinensis* have been found to have high nitrogen activity (Singh, 1993).

7.3. Direct Seeding

The absence of seed rain from a variety of species is one of the primary reasons that impede or obstruct the natural regeneration process in heavily humanized settings. The delayed seed distribution across damaged landscapes restricts the rate of natural healing. The apparent approach to quickening this process is to purposefully reinsert the seed. For direct seeding, the seed has been disseminated, broadcast by land, or broadcast from an aircraft. Typically, the seed needs to be planted on bare ground to germinate quickly and grow into a weed tree. Reintroducing seeds to a mine site before weeds have a chance to establish themselves has been a common practice in mine site rehabilitation initiatives after mining has ended. Proper derelict spot seeding outperformed base root planting in regenerating oaks, pine, black walnut, and black locust. Baumhauer et al., (2005) proposed direct sowing as a technical method for restoring the Forstamt Weissenhorn, the Bavarian region's deteriorated Norway spruce forest. According to the authors, direct seeding seems to be a potential and affordable alternative to planting. Grass, legume, and tree seed mixtures have also been successfully used. Direct sowing of green manures, as described by Martins (2018), is one method for reducing agrochemical expenditures and use in restoration areas.

Direct seeding has the benefit of being inexpensive; there is no need to grow seedlings in nurseries, and they can be dispersed across the landscape easily. There are several disadvantages too. One is that it requires a large volume of the seed of the desired species because only a small proportion of the seed broadcast is usually able to germinate and thrive. Some seeds will be lost to seed predators, some will fail to germinate under field conditions, and some seedlings will die because of dry weather soon after germination.

7.4. Scattered Tree Plantings

To accelerate forest recovery, another way is to foster the structural complexity that attracts seed—or fruit—dispersing fauna into the degraded landscape from a nearly intact forest. One method involves planting small numbers of scattered, single trees or clumps or rows of trees, which form perches for birds. Seedlings are produced from seed sheds below the perch trees. Eventually, the clusters of seedlings grow up to form trees and become bird perches themselves. The clumps of trees enlarge, and the process continues. The trees initially planted might bring more species with seeds not dispersed by animals (e.g., species with large fruit, seeds, or wind-dispersed species) or those where fruiting only occurred infrequently. The approach is likely to be most useful in abandoned far lands with grasslands or shrubs and in locations with few trees (Singh, 1993).

The advantage of this method is that it is comparatively inexpensive because large numbers of seedlings are not needed. The disadvantage is that it is only applicable where there is enough wildlife to cross degraded lands because the majority of plant species are dispersed by wildlife, while only a few species are dispersed by wind. The revegetation rate beneath scattered trees is likely to be slow since many newly germinated seedlings are already covered with grass and weeds.

7.5. Framework Species Method

In this approach, we are planting a small number of "nurse tree" species that can attract seed-dispersing birds. This approach was discussed by Elliott et al., (2003). There are 1000 or even more species planted per ha, which means treatment costs are higher, but it can be avoided by using a smaller number of species. Such species are chosen that can tolerate the site conditions and are attractive to wildlife, so that they are able to reproduce quickly and spread across the site.

The advantage of this approach is that once the trees are established, they soon outcompete grass and weeds, making it easier for the species brought in by seed-dispersing animals to become established. The approach is especially suited to areas close to intact forests that can act as a source of seeds and wildlife; this allows additional species to be recruited quickly.

7.6. Seed Bank Manipulation

A soil seed bank, also known as a seed reservoir, is a collection of ungerminated seeds that may be capable of replacing adult plants after their death and the decay of vegetation (Srivastava, 2022). Rather than newly arrived seeds through dispersal, seed banks in recently eliminated forests play an important role in regeneration. Seed and seedling characteristics differ depending on climate and forest type. Knowledge of germination and growth of seedlings is important for the recruitment of plants, the succession of communities, and the restoration of the forest ecosystem (Khurana & Singh, 2001).

The use of a soil seed bank in restoration decreases over time due to seed viability loss, particularly of the desired species. As my example spoils, this technique offers a tremendous opportunity to quickly re-vegetate. In this, the removal of the top 5 cm of soil from an area covered with vegetation and placing it above the mine spoil gives a good result. Correct selection of donor soil will depend on prior knowledge of the species composition of the seed bank, the time of collection of donor soil, the seed viability of the donor soil, etc.

7.7. Restoration Acceleration by Biological or Fertilizer Treatment

Inoculation of mycorrhizal, blue-green algae, or plant growth-promoting bacteria in degraded ecosystem topsoil can be useful in the re-vegetation of such disturbed sites (Singh & Srivastava, 2022). It has been reported that after mycorrhizal infection, seeding establishments under stressed conditions due to a lack of water and nutrients increased. Usually, it has been found that in most disturbed forests, available nutrients are deficient. Under such conditions, forest recovery can be accelerated by the addition of fertilizers (Singh et al., 1989). It was found that there is a significant increase in the height of seedlings of some tree species following application of fertilizer (urea, NPK). It was reported that adding nutrients to ecosystems can be expected to increase productivity, biomass, and dominance. But it decreases diversity. To increase diversity, it is necessary to increase habitat heterogeneity by using fertilized and unfertilized strips alternately.

For successful large-scale restoration, one should keep the following in mind: (1) learning from the past; (2) ecological knowledge supporting forest restoration; (3) advances in restoration and regeneration techniques and systems; (4) forest restoration following biotic and abiotic disturbances; and (5) restoring forest landscapes for the future.

Conclusion

Forest restoration reduced, and in some cases reversed, biodiversity loss. It also mitigates climate change and fulfills local needs. There are many ways and methods for forest restoration; however, the best method is to enhance structural complexity by planting species and making ecosystems function intact.

References

Aerts, R., Honnay, O. (2011). Forest restoration, biodiversity and ecosystem functioning. *BMC Ecolology* 11, 29. https://doi.org/10.1186/1472-6785-11-29.

Baland, J. M., Bardhan, P., Das, S., & Mookherjee, D. (2010). Forests to the people: Decentralization and forest degradation in the Indian Himalayas. *World Development, 38*(11), 1642-1656.

Baumhauer, H., Madsen, P., Santurf, J. A. (2005). Regeneration by direct seeding: A way to reduce costs of conversion. pp. 359-354. In: Stanturf JA, Madsen P, editors. *Restoration of Boreal and Temperate Forests*. Boca Raton: CRC Press.

Burney, O. T., & Jacobs, D. F. (2018). Species selection –A fundamental silvicultural tool to promote forest regeneration under high animal browsing pressure. *Forest Ecology and Management, 408*, 67-74.

Cernansky, R. (2018). How to plant a trillion trees. *Nature, 560*(7720), 542-544.

Chazdon, R. L. (2008). Beyond deforestation: restoring forests and ecosystem services on degraded lands. *Science, 320*(5882), 1458-1460.

Cortina, J., Amat, B., Castillo, V., Fuentes, D., Maestre, F., Padilla, F., & Rojo, L. (2011). The restoration of vegetation cover in the semi-arid Iberian southeast. *Journal of Arid Environments, 75*(12), 1377-1384.

Crouzeilles, R., Ferreira, M. S., Chazdon, R. L., Lindenmayer, D. B., Sansevero, J. B., Monteiro, L., Strassburg, B. B. (2017). Ecological restoration success is higher for natural regeneration than for active restoration in tropical forests. *Science advances, 3*(11), e1701345.

Devi, U., & Behera, N. (2003). Assessment of plant diversity in response to forest degradation in a tropical dry deciduous forest of Eastern Ghats in Orissa. *Journal of Tropical Forest Science*, 147-163.

Dumroese, K. R., Landis, T. D., Pinto, J. R., Haase, D. L., Wilkinson, K. W., & Davis, A. S. (2016). Meeting forest restoration challenges: using the target plant concept. *Reforesta, 1*(1), 37-52.

Elliott, S., Navakitbumrung, P., Kuarak, C., Zankum, S., Anusarnsunthorn, V., Blakesley, D. (2003). Selecting framework tree species for restoring seasonally dry tropical forests in northern Thailand based on field performance. *Forest Ecology and Management* 184,177-191.

FAO. (2010) *Planted forests in sustainable forest management: A statement of principles*. FAO, Rome, Italy.

FAO. (2011). *Assessing Forest Degradation: Towards the Development of Globally Applicable Guidelines, Forest Resources Assessment*, FAO, Rome, Italy.

FAO. (2013). *Towards the Assessment of Trees Outside Forests*. FAO, Rome, Italy.

FAO. (2018). *The State of the World's Forests 2018 - Forest pathways to sustainable development.*

FAO. (2020) *Forest Resources Assessment 2020 – Key findings*. Rome, Italy.

FAO. (2022). *The State of the World's Forests 2022. Forest pathways for green recovery and building.*

Felton, A. M., Felton, A., Raubenheimer, D., Simpson, S. J., Krizsan, S. J., Hedwall, P. O., & Stolter, C. (2016). The nutritional balancing act of a large herbivore: an experiment with captive moose (Alces alces L). *Plos one, 11*(3), e0150870.

Ghazoul, J., Burivalova, Z., Garcia-Ulloa, J., & King, L. A. (2015). Conceptualizing forest degradation. *Trends in ecology & evolution, 30*(10), 622-632.

Gómez-Aparicio, L. (2009). The role of plant interactions in the restoration of degraded ecosystems: a meta-analysis across life-forms and ecosystems. *Journal of Ecology, 97*(6), 1202-1214.

Gómez-Aparicio, L., Zamora, R., Gómez, J. M., Hódar, J. A., Castro, J., & Baraza, E. (2004). Applying plant facilitation to forest restoration: a meta-analysis of the use of shrubs as nurse plants. *Ecological applications, 14*(4), 1128-1138.

Grossnickle, S. C., & El-Kassaby, Y. A. (2015). Bareroot versus container stocktypes: a performance comparison. *New Forests, 47*(1), 1-51.

Haase, D. L., & Davis, A. S. (2017). Developing and supporting quality nursery facilities and staff are necessary to meet global forest and landscape restoration needs. *Reforesta*(4), 69-93.

He, Q., Bertness, M. D., & Altieri, A. H. (2013). Global shifts towards positive species interactions with increasing environmental stress. *Ecology letters, 16*(5), 695-706.

Holl, K. D., Loik, M. E., Lin, E. H., & Samuels, I. A. (2000). Tropical montane forest restoration in Costa Rica: overcoming barriers to dispersal and establishment. *Restoration ecology, 8*(4), 339-349.

IITO. (2002). *ITTO guidelines for the restoration, management and rehabilitation of degraded and secondary tropical forests*. Yokohama, JP: International Tropical Timber Organisation, IUCN.

Jacobs, D. F., Dalgleish, H. J., & Nelson, C. D. (2013). A conceptual framework for restoration of threatened plants: the effective model of American chestnut (Castanea dentata) reintroduction. *New Phytologist, 197*(2), 378-393.

Jacobs, D. F., Oliet, J. A., Aronson, J., Bolte, A., Bullock, J. M., Donoso, P. J., Rey-Benayas, J. M. (2015). Restoring forests: what constitutes success in the twenty-first century? *New Forests, 46*(5), 601-614.

Khurana, E. & Singh, J. S. (2001). Ecology of tree seed and seedlings: Implications for tropical forest conservation and restoration. Current Science 80(6), 748-757.

Kuijper, D. P., Jędrzejewska, B., Brzeziecki, B., Churski, M., Jędrzejewski, W., & Żybura, H. (2010). Fluctuating ungulate density shapes tree recruitment in natural stands of the Białowieża Primeval Forest, Poland. *Journal of Vegetation Science, 21*(6), 1082-1098.

Lamb, D., & Gilmour, D. (2003). *Rehabilitation and restoration of degraded forests*. Gland, Switzerland and Cambridge, UK and WWF, Gland, Switzerland: IUCN-The World Conservation Union.

Lindenmayer, D. B. (2019). Integrating forest biodiversity conservation and restoration ecology principles to recover natural forest ecosystems. *New Forests, 50*(2), 169-181.

Lindenmayer, D. B., Steffen, W., Burbidge, A. A., Hughes, L., Kitching, R. L., Musgrave, W., Werner, P. A. (2010). Conservation strategies in response to rapid climate change: Australia as a case study. *Biological Conservation, 143*(7), 1587-1593.

Löf, M., Dey, D. C., Navarro, R. M., & Jacobs, D. F. (2012). Mechanical site preparation for forest restoration. *New Forests, 43*(5), 825-848.

Löf, M., Madsen, P., Metslaid, M., Witzell, J., & Jacobs, D. F. (2019). Restoring forests: regeneration and ecosystem function for the future. *New Forests, 50*(2), 139-151.

Lund, H. G. (2009). What is a degraded forest. *Gainesville, VA. USA: Forest Information Services.*

Macdonald, S. E., Landhäusser, S. M., Skousen, J., Franklin, J., Frouz, J., Hall, S., Quideau, S. (2015). Forest restoration following surface mining disturbance: challenges and solutions. *New Forests, 46*(5), 703-732.

Maltoni, A., Mariotti, B., Tani, A., Martini, S., Jacobs, D. F., & Tognetti, R. (2019). Natural regeneration of Pinus pinaster facilitates Quercus ilex survival and growth under severe deer browsing pressure. *Forest Ecology and Management, 432*, 356-364.

Martins, S. V. (2018). Alternative Forest Restoration Techniques. *New Perspectives in Forest Science*. https://www.intechopen.com/chapters/58714.

Meli, P., Holl, K. D., Rey Benayas, J. M., Jones, H. P., Jones, P. C., Montoya, D., & Moreno Mateos, D. (2017). A global review of past land use, climate, and active vs. passive restoration effects on forest recovery. *Plos one, 12*(2), e0171368.

Metslaid, M., Palli, T., Randveer, T., Sims, A., Jogiste, K., & Stanturf, J. (2014). The condition of Scots pine stands in Lahemaa National Park, Estonia 25 years after browsing by moose (Alces alces). *Boreal Environment Research 18 (A): 25-34, 18*(A), 25-34.

Minnemayer, S., Laestadius, L., Sizer, N., Saint-Laurent, C., Popapov, P. (2011). A world of opportunity. *World Resource Institute*, Report, Washington, DC.

Modica, G., Merlino, A., Solano, F., & Mercurio, R. (2015). An index for the assessment of degraded Mediterranean forest ecosystems. *Forest systems, 24*(3), 5.

Morales-Barquero, L., Skutsch, M., Jardel-Peláez, E. J., Ghilardi, A., Kleinn, C., & Healey, J. R. (2014). Operationalizing the definition of forest degradation for REDD+, with application to Mexico. *Forests, 5*(7), 1653-1681.

Oliet, J. A., & Jacobs, D. F. (2012). Restoring forests: advances in techniques and theory. *New Forests, 43*(5), 535-541.

Oliet, J. A., Blasco, R., Valenzuela, P., Melero de Blas, M., & Puértolas, J. (2019). Should we use meshes or solid tube shelters when planting in Mediterranean semiarid environments? *New Forests, 50*(2), 267-282.

Potter, K. M., Jetton, R. M., Bower, A., Jacobs, D. F., Man, G., Hipkins, V. D., & Westwood, M. (2017). Banking on the future: progress, challenges and opportunities for the genetic conservation of forest trees. *New Forests, 48*(2), 153-180.

Puettmann, K. J., Ares, A., Burton, J. I., & Dodson, E. K. (2016). Forest restoration using variable density thinning: Lessons from Douglas-fir stands in western Oregon. *Forests, 7*(12), 310.

Reid, J. L., Fagan, M. E., & Zahawi, R. A. (2018). Positive site selection bias in meta-analyses comparing natural regeneration to active forest restoration. *Science advances, 4*(5), eaas9143.

Rose, K. M. E., Baribault, T. W., & Jacobs, D. F. (2016). Alternative field fertilization techniques to promote restoration of leguminous Acacia koa on contrasting tropical sites. *Forest Ecology and Management, 376*, 126-134. doi: https://doi.org/10.1016/j.foreco.2016.06.012.

Sasaki, N., Asner, G. P., Knorr, W., Durst, P. B., Priyadi, H. R., & Putz, F. E. (2011). Approaches to classifying and restoring degraded tropical forests for the anticipated

REDD+ climate change mitigation mechanism. *Iforest-Biogeosciences and Forestry, 4*(1), 1.

Shono, K., E. A. Cadaweng and P. B. Durst (2007). Application of Assisted Natural Regeneration to Restore Degraded Tropical Forestlands. *Restoration Ecology*, 15(4), 620–626.

Silva, L. N., Freer-Smith, P., & Madsen, P. (2019). Production, restoration, mitigation: a new generation of plantations. *New Forests, 50*(2), 153-168. https://link.springer.com/article/10.1007/s11056-018-9644-6.

Singh A. & Srivastava R. (2022). Impact of Plant growth promoting Rhizobacteria on restoring soil and crop attributes. *Indian Journal of Agricultural Sciences.*

Singh J. S. (1993). *Restoration of degraded land: Concept and strategies, Rastogi publication*, Meerut.

Singh, J. S., Raghubanshi, A. S., Singh, R. S., & Srivastava, S. C. (1989). Microbial biomass acts as a source of plant nutrients in dry tropical forest and savanna. *Nature, 338*, 499–500.

Srivastava, R. & Singh K. P. (2013). Implications of multipurpose tree leaf application on wheat productivity in dry tropics. *Journal of Forestry Research* 24(4), 777-782.

Srivastava, R. & Singh K. P. (2014). Variation in wheat yield, microbial biomass and N-availability in tropical dryland agroecosystem: Impact of application of different tree leaves. *International Journal of Current Microbiology and Applied Sciences* 3(4): 830-842.

Srivastava, R. (2022). *Variations in weeds and seed bank diversity in different ecosystem: A comparative study of two agroecosystems and grassland ecosystem in Indian dry tropical region (Editor:* Srivastava R.)-LAP Lambert Academic publishing, 150p.

Stanturf, J. A., Palik, B. J., Williams, M. I., Dumroese, R. K., & Madsen, P. (2014). Forest restoration paradigms. *Journal of sustainable forestry, 33*(sup1), S161-S194.

Stanturf, J. A., Schoenholtz, S. H., Schweitzer, C. J., & Shepard, J. P. (2001). Achieving restoration success: myths in bottomland hardwood forests. *Restoration ecology, 9*(2), 189-200.

Thompson, I. D., Ferreira, J., Gardner, T., Guariguata, M., Koh, L. P., Okabe, K., Barlow, J. (2012). Forest biodiversity, carbon and other ecosystem services: relationships and impacts of deforestation and forest degradation. *IUFRO World Series Volume 31. p. 21-51, 31*, 21-50.

Thompson, I. D., Guariguata, M. R., Okabe, K., Bahamondez, C., Nasi, R., Heymell, V., & Sabogal, C. (2013). An operational framework for defining and monitoring forest degradation. *Ecology and Society, 18*(2).

Thompson, I. D., Okabe, K., Tylianakis, J. M., Kumar, P., Brockerhoff, E. G., Schellhorn, N. A., Nasi, R. (2011). Forest biodiversity and the delivery of ecosystem goods and services: translating science into policy. *BioScience, 61*(12), 972-981.

Thompson, I., Mackey, B., McNulty, S., & Mosseler, A. (2009). Forest resilience, biodiversity, and climate change. *Paper presented at the Secretariat of the Convention on Biological Diversity*, Montreal. Technical Series no. 43. 1-67.

Tucker, C. M., Randolph, J. C., Evans, T., Andersson, K. P., Persha, L., & Green, G. M. (2008). An approach to assess relative degradation in dissimilar forests: toward a comparative assessment of institutional outcomes. *Ecology and Society, 13*(1).

UN. (2019). Press release New UN decade on ecosystem restoration offers unparalleled opportunity for job creation, food security and addressing climate change. *UNEP*, 2019-03-01, https://www.unep.org.

Villar-Salvador, P., Uscola, M., & Jacobs, D. F. (2015). The role of stored carbohydrates and nitrogen in the growth and stress tolerance of planted forest trees. *New Forests, 46*(5), 813-839.

Wagner, R. G., Gonzalez-Benecke, C. A., Nelson, A. S., & Jacobs, D. F. (2018). Forest regeneration in changing environments. *New Forests, 49*(6), 699-703.

Zapico, I., Duque, J. F. M., Bugosh, N., Laronne, J. B., Ortega, A., Molina, A., Castillo, L. S. (2018). Geomorphic reclamation for reestablishment of landform stability at a watershed scale in mined sites: The Alto Tajo Natural Park, Spain. *Ecological Engineering, 111*, 100-116.

Chapter 5

The Role of Microorganisms in Ecosystem Restoration

Nidhi Singh[1]
Rupanshee Srivastava[1]
Tripti Kanda[1]
Sadhana Yadav[1]
Shivam Yadav[2]
and Neelam Atri[3,*]
[1]Department of Botany, Institute of Science, Banaras Hindu University, Varanasi, India
[2]Department of Botany, University of Allahabad, Prayagraj, India
[3]Department of Botany, MMV, Banaras Hindu University, Varanasi, India

Abstract

The exploitation of natural resources is still the greatest issue to deal with environmental problems in the age of science and technology. As a direct response to the deterioration and extinction of ecosystems, ecological restoration is undertaken all over the world. It has huge potential to significantly enhance public health in addition to having an ecological impact. Depending on regional circumstances and social preferences, ecosystem restoration spans a broad continuum of approaches to stopping and reversing deterioration that can lead to better ecosystem services and restore biodiversity. Microorganisms are nature's greatest gift for regenerating lost ecosystems, because over billions of years, despite numerous global changes, microbes have been

* Corresponding Author's Email: neelam1409@bhu.ac.in.

In: Conserving, Halting and Reinstating Degraded Natural Resources …
Editors: Rajani Srivastava and Pankaj Kumar
ISBN: 979-8-89113-067-8

a constant on our globe. The fundamental components that make up all living systems, particularly carbon, oxygen and nitrogen, may be recycled by microorganisms, which is the most significant impact they have on the planet. By decomposing organic waste, microbes serve significant roles in agriculture and the environment. In wastewater treatment plants, microbial communities have been cleaning the water for ages. Recent studies have shown that the same breakdown skills also apply to explosive materials, poisonous compounds, and even plastics. Numerous bacteria can fix carbon or nitrogen, transforming the toxic compounds into forms that can be used. Therefore, microbes play a vital role in sustainable development. This chapter discusses methods and the significance of using microbes to restore ecosystems.

Keywords: ecosystem restoration, UN decade, microorganism, bio-control, sustainable

1. Introduction

Materials circulate through the lithosphere (rocks), atmosphere (air), hydrosphere (water), and biosphere (organisms) on Earth, which is referred to as a "closed system" (Figure 1). An individual or community's environment is made up of a variety of social and cultural factors. The physical environment and a biological community make up an ecosystem. One of the key problems in the current situation is environmental sustainability and protection. Changing global environments are brought about by entering an anomalous era of concurrent buffers, where rates of change in human cultures and ecosystems are protected from cataclysmic collapse by internal diversity that guards the entire ecology against exaggerated, unusual behaviors or hereditary malformations. The ability of human activity to change the ecosystem has become remarkable. The study of restoration ecology and its applied methods are therefore a potentially low-cost, buoyant answer to this rising dilemma (Suding et al., 2015). As our biosphere is home to a variety of microorganisms, therefore, today's ecological restoration is influenced by the interaction between microbes and the environment. As a result, deprived and degraded ecosystems are regenerated through the establishment of flora and fauna, especially on a broad scale, which offers preferences to minimize environmental degradation. The Society for ecological restoration (SER) was founded internationally to promote ecological restoration (Winterhalder et al., 2004) as a means of preserving ecosystem diversity and reestablishing a positive ecological balance.

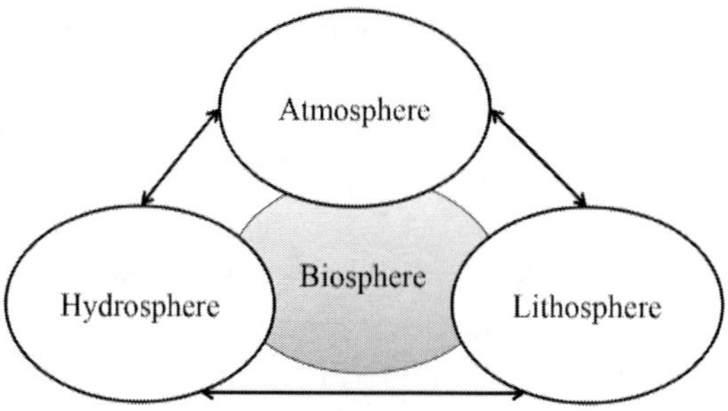

Figure 1. Interactions between different spheres of Earth.

2. Ecosystem Restoration

An ecosystem is a region where different species of plants, animals and other organisms coexist in peace with the environment, including the weather. The ecology of ecosystems focuses on the interaction between living things and their surroundings as a comprehensive system. Environmental deterioration is a grave global problem. The Industrial Revolution, the population boom and the increased desire for luxury goods in daily life are the primary causes of the issue. The state of nature and its resources is currently deteriorating due to a lack of sufficient education, awareness, understanding and attitude among people toward the environment. The unique feature of the current environmental issues is that they are primarily driven by anthropogenic rather than natural processes. Ecological restoration entails restoring the habitat and ecosystem services of the lands and rivers that support the ecosystems that support the plants and animals. One of the most crucial methods for providing biologically based solutions to issues like food insecurity, climate change mitigation and adaptation, and biodiversity loss is ecosystem restoration. The United Nations General Assembly proclaimed “2021-2030” as the United Nations Decade on Ecosystem Restoration (hereafter the “UN Decade”), by enacting Resolution 73/284, the United Nations has recognized the urgent need to halt, prevent and reverse ecosystem degradation and to restore degraded terrestrial, freshwater and marine ecosystems worldwide. It is necessary to have a common understanding of ecosystem restoration to aid in the UN Decade’s implementation and assist it accomplish its objectives, defined as "the procedure for stopping and reversing deterioration, leading to better ecosystem services and restored biodiversity. Depending on regional factors and human preferences, ecosystems cover a broad spectrum of techniques" (Dockendorff et al., 2022).

According to the UN Decade, there are 10 key ideas for restoring the ecosystem:

1. Restoration of ecosystems supports both the RIO conventions and the UN's Sustainable development goals - Achieving ambitious global targets for supporting life on Earth requires the implementation of restoration projects, programs and initiatives at all spatial dimensions from local sites to vast landscapes and seascapes.
2. Ecosystem restoration encourages inclusive and participative governance, social justice and equity from the beginning to the end of the process and the results - Such broad-based engagement is essential for attaining the long-term restoration goals and it should be encouraged as much as possible at every stage of the process, from planning to monitoring. This participation can be attained by ensuring equal and consistent access to knowledge and information, acknowledging and addressing social inequalities through the empowerment and capacity development of underrepresented groups, seeking free, prior and informed consent, offering strong incentives and enhancing the livelihoods, food security and opportunities for local communities, promoting co-management and ensuring a crucial role for local communities in decision-making and acknowledging and addressing social inequalities.
3. Ecosystem restoration includes a continuum of restorative activities- The goal of ecosystem restoration is to restore all types of damaged ecosystems by using a variety of activities, either individually or together. However, for an activity to be termed ecosystem restoration, it must enhance human well-being, ecological health and integrity, biodiversity and sustainably producing commodities and services. Ecosystem restoration is a practice that can be applied to all forms of degraded ecosystems, landscapes and seascapes, including urban, agricultural, cultural, semi-natural and natural systems. Major types of restorative exercises (Cohen-Shacham et al., 2019) include

a) Decrease detrimental effects on the environment and society, including pollution and the use & management of unsustainable resources.
b) Remediation is the process of removing toxins, pollutants and other contaminants.
c) Restoration of ecosystem functions and services in heavily altered locations such as former mining sites and damaged manufacturing systems.
d) Ecological restoration, which takes into account environmental change and strives to reverse ecological degradation, helps return an ecosystem to the course it would have taken in the absence of the deterioration.

4. Biological diversity, ecosystem health and integrity, and human well-being are goals of ecosystem restoration- Given project- and program-level objectives, ecosystem restoration attempts to produce and sustain the highest net gain for biodiversity, ecosystem health and integrity, ecosystem products and services, climate change mitigation and human health and well-being at local, national and global scales. Restorative management techniques should aid and support the natural processes of recovery rather than causing more deterioration (Andrade et al., 2011). It is preferable to use the genetically acceptable germplasm of native species. While using non-native species that may or may already be invasive should be avoided.
5. The primary and indirect causes of ecological deterioration are addressed via ecosystem restoration- The loss of biodiversity, ecosystem products and services, and the direct and indirect causes of ecosystem degradation and fragmentation should all be simultaneously addressed in all restorative activities. Restorative efforts may ultimately fail if the issues are not addressed. The degree and causes of deterioration should be determined during the planning phase of restoration projects, programs, or initiatives. Then, steps should be created to lessen and mitigate their consequences at the appropriate scale.
6. All sorts of knowledge are incorporated into ecosystem restoration, and their integration and interchange are encouraged at every stage of the procedure. All sorts of knowledge are incorporated into ecosystem restoration, and their integration and interchange are

encouraged at every stage of the procedure to attain better kinship with nature, cooperation and efficacy, ecosystem restoration should work to incorporate all types of knowledge and practices, including but not limited to Indigenous, traditional, local, and scientific methods of knowing. Platforms and networks for storing, integrating, and disseminating that knowledge and information ought to be created, and channels for communication and dissemination should be regularly updated, simple to use, clear to understand and sensitive to cultural contexts, to facilitate the exchange of knowledge and information (taking into account languages and literacy levels).

7. To restore an ecosystem, clear short-, medium-, and long-term ecological, cultural, and socioeconomic objectives and goals must be established- This should include targets and indicators that are measurable against the baseline condition, and that specify the direction (e.g., rise or decline) and size of the desired change and are time-bound, where appropriate. Ecological, cultural, and socioeconomic trade-offs should be addressed and resolved through open, honest negotiation in a way that doesn't jeopardize ecosystem recovery.
8. Ecosystem restoration is adapted to the local ecological, cultural, and economic circumstances while taking the greater landscape or seascape into account- Taking into account threats from the greater landscape or seascape, exchanges of energy and organisms across ecosystem boundaries, ecological and hydrological connections, as well as trans-boundary effects, is essential for effective restoration. Utilizing spatial planning techniques will make it easier to adapt projects, programs and initiatives to the larger landscape, seascape, or eco-region to maximize gains in terms of biodiversity, ecological health and integrity, and human well-being, including sustainable production of products and services (Sayer et al., 2013).
9. Monitoring, assessment, and adaptive management are all a part of ecosystem restoration, both during and after the project's existence- At local, national, and international levels, the participation of stakeholders in monitoring can foster social learning, the building of capacity and communication between stakeholder groups and communities of practice. Given that restoration is a long-term project and that conditions will change, adaptive management, which is an iterative process of monitoring, evaluating, reflecting

and adapting activities and approaches as necessary, enables the identification of unexpected (both positive and negative) outcomes and improves future actions. To capture medium- and longer-term impacts, monitoring should continue after the project, program, or initiative has ended.

10. Policies and strategies that support ecosystem restoration's long-term development, enabling replication and scaling-up, are necessary. In order to maximise the long-term net gain from restorative initiatives, it is also necessary to: coordinate operations among institutions, sectors and stakeholders through an effective governance system; Fostering political commitment on a local, national and international level as well as trans-boundary agreements; presenting opportunities for capacity-building to enable the individuals, groups, institutions and networks engaged in restoration; mainstreaming successful practices to have widespread impact and permit replication; locating, securing, and sustaining sufficient funding (from the public sector, the private sector, international organisations or other sources) to complete all phases of the process; creating revenue streams (for instance, through sustainable production, ecotourism, payment for ecosystem services and other sustainable uses of natural resources) that do not jeopardise the integrity of the restoration process and support its viability; and ensuring the safety of participants and right holders, particularly in regions of political conflict or conflict over natural resources (Manchikanti, 2016).

2.1. Ecosystem Restoration Is Now a Global Challenge and Priority

Ecosystem restoration is now globally recognized as a crucial part of conservation efforts and as is necessary for the long-term viability of our human-dominated planet. Restoration scientists and practitioners are already on the frontlines and will increasingly be asked to participate in large–scale programs addressing the current environmental crises and problems. The fourth world congress of the International Union for the Conservation of Nature (IUCN) brought together nearly 10,000 representatives of NGOs and governments in Jeju, Korea in September 2012 and they issued the Jeju Declaration to work toward a future where conservation, restoration and sustainable use enhance biodiversity conservation and the provision of ecosystem services while fostering sustainable development and improved livelihoods (IUCN, 2012). Ecological restoration science and practice have developed quickly in the last decade, giving rise to a variety of information, resources and technologies. Among many recent examples, the IUCN World Commission on protected areas recently published a new volume in Jeju which comprises fundamentals, best methods and recommendations regarding the ecological restoration for protected areas (Gann et al., 2019).

Although there are still many obstacles to overcome in research and practice, the restoration community is well-positioned to play a significant role in aiding the global community in addressing urgent biodiversity and environmental issues that threaten our sustainability and thus human well-being (Figure 2).

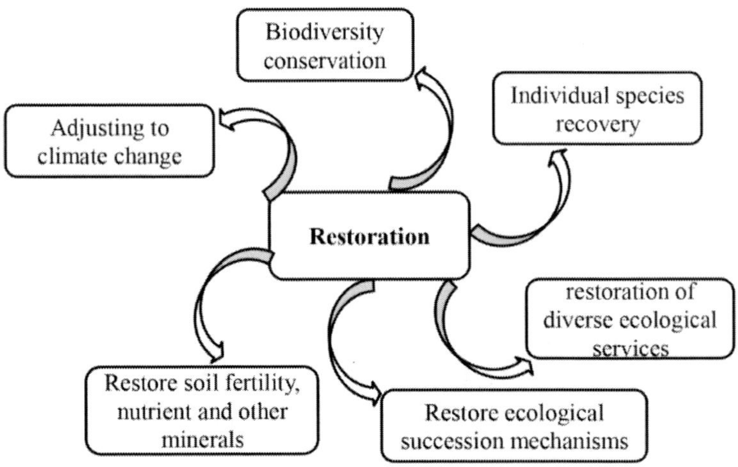

Figure 2. Advantages of ecological restoration.

For designing and implementing knowledge-based, cost-effective and participatory restoration projects and programs, specific tasks are required

1. Capacity-building activities, such as local workshops and technical training programs.
2. Sharing of knowledge via searchable database, which includes e-learning courses, case studies and best practices.
3. Exchange initiatives between organizations, restoration professionals and researchers.
4. Restoration of the ecology into more general planning procedures.

5. Raising awareness among the public, decision-makers and environmental managers of the economic, ecological and social benefits of ecosystem restoration.

It is crucial to understand each factor that leads to ecological deterioration before studying any information about ecological restoration. Because of the increasing human population, environmental breakdown is caused by the consumption of resources such as water, air and soil (Singh et al., 2019). Human activities involve utilizing energy present in the biosphere and excessive use of this energy leads to uneven environmental degradation. Activities that give rise to ecological destruction are soil decrement, vegetation loss, soil alteration with salt and dust, air pollution, water pollution, etc. Therefore, maintaining a balanced species relationship can maximize ecosystem benefits and help to offset the effect of a changing climate, both of which may have a mutual impact on overall sustainable development, either directly or indirectly. Despite the benefits of a consistent framework for tracking and reporting ecosystem restoration activities, issues are sure to arise during the development and eventual implementation of restoration reporting frameworks. These obstacles fall into three categories: technological, institutional, and financial. When the framework is created, considerations must be made to ensure that its application meets user expectations and produces the desired outcomes.

3. Roles of Micro-Organisms in Ecosystem Restoration

Microorganisms are ubiquitous; these are the organisms that are economically, socially and environmentally important. They play a crucial part and serve as the primary engineers in controlling all ecological processes. They bring about ecological shifts and serve as a universal trigger. They are everywhere and can carry out any desired action. The biotic basis for all life is provided by microorganisms. In the presence of a nucleus, microbes were categorized into two groups, prokaryotes and eukaryotes. Protists, fungi, algae, and other microbes with nuclei in cells are classified as eukaryotes, while bacteria, viruses, and related archaea are prokaryotes. (Table 1). They aid in preserving atmospheric nitrogen and carbon dioxide, which are currently being depleted by greenhouse gases and other damaging things. Bacteria produce about 50% of the total oxygen and help in the production and fixation of nitrogen. Approximately, 70% of the methane is produced by bacteria. Certain microbes also have the potential of producing phytohormones, amino acids, polyamines and biofuels (Figure 3).

Table 1. Major groups of microbes and their ecological roles

S. N.	Microorganism	Ecological Role	Example
1.	Archaea	Environment and atmospheric condition regulation, nutrient cycling.	Thermosphaera aggregans, Ignisphaera aggregans, Sulfolobus tokodaii, Metallosphaera sedula
2.	Chemoautotrophic bacteria	Climate regulation, nutrient cycling, water purification.	Nitrobacter, Nitrosomonas, Sulphur bacteria.
3.	Heterotrophic bacteria	Climate regulation, nutrient cycling, water purification, decomposition.	Agrobacterium, Xanthomonas, Pseudomonas, Salmonella, Rhizobium
4.	Phototrophic bacteria	Synthesis of chlorophyll.	Cyanobacteria, Protoeobacteria
5.	Fungi	Nutrient cycling, soil formation.	Saccharomyces *cerevisiae,* Trichoderma sp., Chaetomium sp.
6.	Arbuscular mycorrhizal fungi	Indirectly affects the primary product.	Glomus mosseae, G. fasciculatum, Gigaspora gilmorei
7.	Protozoans	Nutrient cycling, mineralization, Decomposition.	Agrobacterium, Pseudomonas, Xanthomonas
8.	Viruses	Nutrient cycling.	Bacteriophage

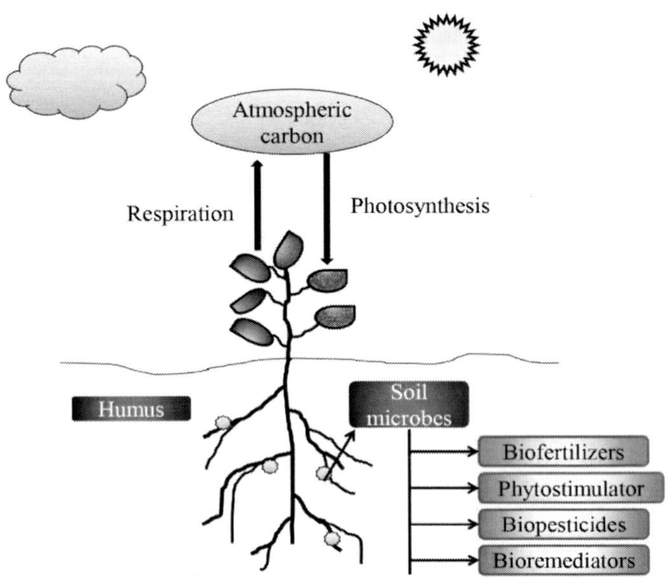

Figure 3. Role of Microorganisms in Plant and Soil.

3.1. Roles of Bacteria and Cyanobacteria

Bacteria are tiny single-celled organisms and found in a variety of environments in forest soils, including the bulk soil, rhizosphere, litter and deadwood habitats, where the availability of nutrients and biotic interactions influence the composition of bacterial populations.

Bacteria help in ecosystem restoration through the following characteristics.

3.1.1. Plant Growth-Promoting Rhizobacteria

Free-living soil microorganisms reside on the roots of some plants and promote plant growth. Additionally, they are essential in combating diseases brought on by nematodes, fungi, viruses, and bacteria. PGPR also assists in coping with environmental stresses. When it comes to preserving soil fertility, PGPR is the main player. Direct stimulation because of PGPR involve the production of phytohormones, biological nitrogen fixation (e.g., *Rhizobium*, *Sinorhizobium, Bradyrhizobium*, *Mesorhizobium*), solubilize minerals like irons and phosphates (e.g., *Azotobacter chroococcum*, *Bacillus circulans*, *Enterobacter agglomerans, Pseudomonas chlororaphis, Pseudomonas*, *Rhizobium*, *Serratia*).

Insoluble organic phosphates can be mobilized and solubilized by cyanobacteria, which increases the bioavailability of phosphorus to plants, and the production of enzymes and siderophores. Meanwhile, indirect stimulation involves the development of antibiotics, the chelation of iron in the rhizosphere, the struggle for rhizosphere niches, and the production of extracellular enzymes to break down the fungus cell wall. *Bacillus subtilis* and *Pseudomonas fluorescens* are the strains of PGPR that are most frequently examined as the most promising for indirect stimulation. (Beattie, 2007).

3.1.2. Nitrogen-Fixing Bacteria and Cyanobacteria

The purpose of nitrogen-fixing bacteria (*Azatobacter, Bacillus, Clostridium*) is to provide plants with a crucial nutrient that they are unable to receive naturally from the air. Nitrogen is essential for the growth of healthy plants since it is one of the primary components of the chlorophyll molecule, which is involved in photosynthesis. In roots, the presence of nitrogen helps in the uptake of water and nutrients. Overall, the soil's nitrogen content has a direct impact on how quickly plants develop and how much protein they produce. Different morphological domains of cyanobacteria can fix atmospheric nitrogen because of heterocyst formation. Fixed nitrogen by cyanobacteria is accumulated by the glutamine synthetase–glutamate synthase (GS-GOGAT) pathway (Miflin & Lea, 1976).

Total contributions toward the overall nitrogen content by prokaryotes were mainly controlled by different biotic and physiochemical factors. Microorganisms for fixing nitrogen require 16mol of adenosine triphosphate (ATP) to reduce each mole of Nitrogen. Non-photosynthetic microbes depend on other organisms for getting ATP, whereas, cyanobacteria (*Anabaena cylindrical, A. variabilis)* get their ATP by reduction of Sugars as they are photosynthetic organisms. Another challenge for the bacteria is that oxygen destroys one of the enzymes required for nitrogen-fixing (necessary for efficient ATP formation). Forming symbiotic connections with other creatures that may produce carbohydrates, such as diatoms, the fungi of some lichens, shipworms, termites, and some plants, particularly in nodules of the roots, is one way to solve this issue (Keshani et al., 2021).

3.1.3. Bacterial and Cyanobacterial-Mediated Organic Matter Decomposition

Bacteria recycle garbage from nature rather than just eating it. Bacteria recycle organic or dead stuff to provide the required nutrients to the organisms by breaking them down. On a dry weight basis, bacteria have similar amounts of carbon, oxygen, nitrogen, hydrogen, phosphorus, and sulfur as other cells (55%, 20%, 10%, 8%, 3%, and 1% respectively). In comparison to algae, bacteria have a higher C: N:P ratio (106:19:6), meaning they contain more nutrients. These C: N:P ratios suggest that to obtain the necessary amounts of N and P, bacteria would need to break down more C in plants. By decomposing, it encourages water retention and shields roots from parasites and diseases. It decomposes organic matter toward humus that helps in maintaining fertility and soil health (Keshani et al., 2021).

3.1.4. Bacterial and Cyanobacterial-Mediated Bioremediation

Degradation of toxic compounds through bacteria and cyanobacteria is one of the safest, most cost-effective and eco-friendly methods (Kumar et al., 2017). For the degradation of oil spills on soils/oceans, *Pseudomonas* is one of the well-known bacteria. The two ways to achieve the technology for bioremediation treatment (Figure 4) are given below.

a) In-situ bioremediation- The term "in-situ methodology" refers to the use of methods to remove dirty soil and water from a location while causing the least amount of disruption.
b) Ex-situ bioremediation- application of methods to contaminated soil and water at the site after removal from a hazardous location by soil extraction or pumping (water). And bio-augmentation is the process of accelerating the breakdown of contaminants in polluted places by introducing microorganisms (Baker & Herson, 1994).

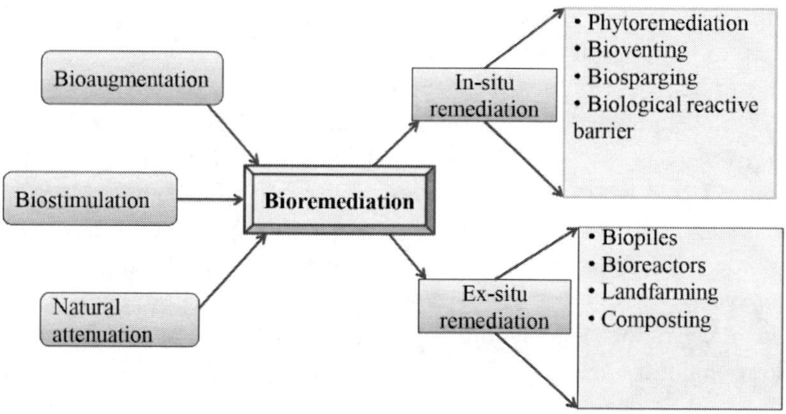

Figure 4. Bioremediation by microorganism.

Table 2. Bioremediation of waste compounds through bacterial species

Waste compounds	Microorganisms	References
Benzene and xylene	Pseudomonas putida	(Safiyanu et al., 2015)
Endosulfan	Bacillus, Staphylococcus	(Mohamed et al., 2011)
Crude oil	Pseudomonas cepacia, Bacillus cereus, Bacillus coagulans, Citrobacter koseri	(Fagbemi & Sanusi, 2016)
Aromatic hydrocarbons	Acinetobacter sp., Microbacterium sp.	(Simarro et al., 2013)
Textile dyes	Micrococcus luteus, Listeria denitrificans, Nocardia atlantica	(Hassan et al., 2013)

In recent years, cyanobacteria have been promoted as experimental organisms in environmental biotechnology programs due to their function in soil fertility, as well as their potential to remove heavy metals from polluted locations and their active involvement in soil reclamation. Cyanobacteria also have the ability to destroy xenobiotics and naturally occurring hydrocarbons found in contaminated areas (e.g., *Cyanospora capsulata* and

Nostoc PCC 7936) (Gupta et al., 2012). Because of their toxicity and risk to both human and animal life as well as the environment, heavy metal contamination is currently a serious problem. Heavy metals (As and Cr) absorption metabolism distinguishes beneficial bacteria such *Alcaligenes faecalis*, *Pseudomonas fluorescens*, and *Enterobacter cloacae* (Sharma, 2021). Utilizing microbial surfactants has shown to be a more recent and effective method of controlling soil and water contamination (Table 2).

3.1.5. Bacterial and Cyanobacterial Surfactants

Complex molecules created by biological processes, such as glycolipids, lipopeptides, lipoproteins, neutral lipids, fatty acids and phospholipids, make up biosurfactants. Microorganisms create surfactants that have the benefit of being biodegradable and extremely specific (Table 3). *Acinetobacter calcoaceticus* is a type of bacteria that aids in binding heavy metals like uranium (Ron & Rosenberg, 2001). Elisngela (Elisngela et al., 2015) investigated the cyanobacteria strains *Arthrospira* sp. LEB 18 and *Synechococcus nidulans* LEB 25 and the chlorophyte strains *Chlorella minutissima* LEB 108, *Chlorella vulgaris* LEB 106 and *Chlorella homosphaera* for the synthesis of biosurfactants in autotrophic and mixotrophic farming.

Table 3. Important bacterial agents used as biosurfactants

Microorganisms	Types of surfactants	References
B. subtilis	Surfactin	(Hassan et al., 2013)
Ustilago maydis, U. zeae	Cellobiose lipids	(Desai & Banat, 1997)
Pseudomonas sp., P. Aeruginosa, Serratia rubidea	Rhamnolipids	(Desai & Banat, 1997)
Candida apicola, C. bombicola, C. lipolytica, C. bogoriensis, Wickerhamiella domercqiae, C. Batistae, Pichia anomala	Sophorose lipids	(Van Bogaert et al., 2007), (Konishi et al., 2008), (Thaniyavarn et al., 2008)
B. licheniformis	Lichenysin A & B	(Mukherjee et al., 2006)

3.1.6. Bacterial and Cyanobacterial-Mediated Wastewater Management

Wastewater management, which is defined as the collection, treatment, and reuse of wastewater, is a crucial strategy for safeguarding water resources. Anaerobic, aerobic, and facultative types of microbes are used to treat wastewater. Anaerobic bacteria (*Propionibacterium, Bacterioides*, or *Fusobacterium*) are utilized to produce methane gas and reduce the amount of sludge that needs to be broken down. They get oxygen by performing the process of fermentation. Another advantage of anaerobic microorganisms employed in sewage treatment is the removal of phosphorus from wastewater. Aerobic bacteria (*Achromobacter*, *Acinetobacter,* or *Desulfovibrio*) use free oxygen for respiration and to convert wastewater contaminants into energy for growth and reproduction. The intermediate zone of bacteria that can switch from anaerobic to aerobic respiration is known as the facultative zone. They are capable of enduring either anaerobic or aerobic conditions (Singh et al., 2019).

Currently, cyanobacteria (*Nostoc muscorum*, *Anabaena variabilis*, *Lyngbya majuscula* and *Oscillatoria salina*) are gaining popularity in wastewater cleanup because of capacities such as, enhancing the oxygen content, bioremediation of heavy metals, reducing the turbidity of waters (Ahmad, 2022).

4. Protists

Protists are single-celled and multicellular eukaryotic organisms. Most protists are photoautotrophic while others are chemoheterotrophic (*Amoeba*) or can be parasitic (*Trypanosoma* protozoa). So, in nature, they perform the role of both producers and consumers. They play a crucial role in food chains. Protists are sensitive to sulfur, ozone, nitrogen, heavy metals, particulate pollutants, etc. means that they are sensitive to environmental change providing the essential requirement to use protists bio-indicators (Singh et al., 2019).

5. Fungi

Therefore, fungi are more closely linked to plants and animals than bacteria or archaea since they are eukaryotes. Fungi are heterotrophic organisms, which means they obtain their food from nonliving organic sources. Fungi can help in halting global challenges even including starvation and climatic changes (McSweeney & Mackie, 2012). Fungi's advantages are:

- Environmental safety

The fungus can be an effective technique for lowering environmental pollutants, which can degrade various environmental contaminants together with pharmaceuticals, personal care items, plastic, and other petroleum-based goods (Table 4). Fungi plays a significant role in the process of mycoremediation (Perelo, 2010), mycoinsecticides, mycoherbicides as well as in mycorrhizal association.

Table 4. Role of fungal species in waste removal

S. No.	Fungi	Pollutants	References
1.	Pleurotus sajor-caju	Heavy metals	(Jibran, 2011)
2.	Pleurotus sp., Pleurotus pulmonarius, Polyporus sp.	Crude oil/petroleum hydrocarbons	(Kristanti et al., 2011), (Njoku, Yussuf et al., 2016)
3.	Pleurotus ostreatus	Green polyethylene (GP) starch-based plastic polymer degradation and plastic degradation	(da Luz et al., 2013; da Luz et al., 2015)
4.	A. bisporus, Calocybe indica, Pleurotus platypus	Cu, zinc, iron, cadmium, nickel, lead	(Prasad & Sachin, 2013)
5.	Flammulina velutipes	Copper ions from aqueous waste	(Luo et al., 2013)
6.	Fomes fasciatus	Copper ions	(Sutherland & Venkobachar, 2010)
7.	Auricularia sp., Schizophyllum, Polyporus sp.	Malachite green dye was degraded	(Garg & Tripathi, 2017)

- Nutrient Cycling

Fungi has the potential of converting nutrients that plants can uptake. Certain decomposer fungi can break down plant and animal debris, thus, supporting the nutrient cycle and increasing their availability in the soil. Nitrogen and Phosphorus are the two main nutrients required for plant growth and productivity. Fungi also stimulates phosphorus mobilization and nitrogen fixation (Richardson et al., 2009).

- Climate governance by carbon-cycle

Through the soil food web fungi play a major role in the carbon cycle. Decomposer fungi release carbon by decomposing dead organic matter while fungi of mycorrhizal association with roots of plant supply more stable stocks of carbon. As fungi with plants represent a procedure called soil carbon sequestration, this process improves soil fertility and also aids in reducing excess carbon in the atmosphere that occurs due to human activities (Mohan et al., 2018).

Fungi improves plant development, survival, reproduction, nutrient aggregation and absorption, endurance to abiotic and biotic stresses and adjustment of plant and animal communities. Altogether, fungi play a vital role in ecological restoration (Brundrett & Abbott, 2002).

6. Viruses

The virus is one of the dominant microorganisms but their abundance varies between habitats. In aquatic food web systems viruses play a crucial role by inducing mortality of other harmful aquatic microbes (Hewson et al., 2010). Bacteriophages are conceivably the most widespread and ancient known species on Earth, recently they are the theme of interest as we shift toward improving food safety and giving people access to natural foods. The use of Bacteriophage could offer a secure, efficient and eco-friendly method for reducing food-borne bacterial pathogens. As bacteriophage kills the target bacterial pathogen without affecting the natural food and microflora (Sulakvelidze, 2013). Some studies regarding the role of viruses in ecosystem restoration are described below-

a) Shiga shield is a phase preparation made up of five lytic bacteriophages that are particularly designed to target pathogenic *Shigella* species that are usually present in polluted water and contaminated food (Soffer et al., 2017).
b) Additionally, a novel strategy for expanding the lytic activity of endolysins toward Gram-negative cells is highlighted. Six isolated bacteriophages (BFSE16, BFSE18, PaDTA1, PaDTA9, PaDTA10 and PaDTA11) were mixed to form an absorbent food pad. This pad was used as biocontrol and helped in food preservation (Soffer et al., 2017).
c) Studies on the spread of cholera discovered a mysterious trait of the behavior of bacteriophage viruses that has significant medical ramifications. These bacteriophages can be altered to target bacteria and are effective against a wide variety of infections as well as in the production of anti-microbial drugs (Rai & Smith, 2016).

Conclusion

Various anthropogenic activities and the non-selective use of chemical fertilizers with the increasing human population have resulted in several environmental hazards such as loss of fertile land, soil infertility, etc. To counteract these and many other issues, cost-effective, eco-friendly and sustainable approaches are required to be discovered and adopted. Ecological restoration ultimately aims to create self-sustaining ecosystems with the required structure, processes, and services on damaged lands. Ecosystem restoration through microorganisms is a very profitable and glamorous option for cleaning, managing and recovering ecosystems. Microbes study explains how nature is restored by natural organisms only. Microorganisms are essential to the structure, nutrient cycling and plant interactions that take place in the soil. These functions are crucial for restoring ecosystem function and biodiversity. Measuring the community reveals the system's condition concerning restoration goals and the success of management initiatives, and manipulating the community holds promise for accelerating the rate at which damaged systems recover. With exceeding research and development, the use of microbes as bio-agents will become a reality and instrumental to fundamental methods that propel stability and restoration of ecosystems. Thus, guide us toward an ideal ecosystem that is sustainable, maintains and improves human health, and produces enough food for the increasing population.

Acknowledgment

The authors would like to thank University Grant Commission (UGC), New Delhi for providing financial assistance.

References

Ahmad, I. Z. (2022). The usage of Cyanobacteria in wastewater treatment: Prospects and limitations. *Letters in Applied Microbiology*, 75(4), 718–730. https://doi.org/10.1111/lam.13587.

Andrade, Angela, Córdoba, Rocio, Dave, Radhika, Girot, Pascal, Herrera-F., Bernal, Munroe, Robert, Oglethorpe, Judy, Paaby, Pia, Pramova, Emilia, Watson, James, and Vergar, Walter (2011). *Draft principles and guidelines for integrating ecosystem-based approaches to adaptation in project and policy design: a discussion document.* Turrialba, Costa Rica: Centro Agronómico Tropical de Investigación y Enseñanza [Tropical Agricultural Research and Teaching Center] (CATIE).

Baker, K. H. & Herson, D. S. (1994). Introduction and overview of bioremediation, in: Baker, K. H. & Herson, D. S. (Eds 3), *Bioremediation*, McGraw-Hill, New York.

Beattie, G. A. (2007). Plant-associated bacteria: survey, molecular phylogeny, genomics and recent advances. In: Gnanamanickam, S. S. (ed.) *Plant associated bacteria*, Chapter 1. Springer, Netherland, pp 1-56.

Brundrett, M. C., & Abbott, L. K. (2002). *Arbuscular mycorrhizas in plant communities. Microorganisms in plant conservation and biodiversity* (pp. 151–193). Springer.

Cohen-Shacham, E., Andrade, A., Dalton, J., Dudley, N., Jones, M., Kumar, C., Maginnis, S., Maynard, S., Nelson, C. R., Renaud, F. G., Welling, R., & Walters, G. (2019). Core principles for successfully implementing and upscaling Nature-based Solutions. *Environmental Science and Policy*, *98*, 20–29. https://doi.org/10.1016/j.envsci.2019.04.014.

Da Luz, J. M. R., Paes, S. A., Nunes, M. D., da Silva, Mde C., & Kasuya, M. C. (2013). Degradation of oxo-biodegradable plastic by Pleurotus ostreatus. *PLOS ONE*, *8*(8), e69386. https://doi.org/10.1371/journal.pone.0069386.

Da Luz, J. M. R., Paes, S. A., Ribeiro, K. V., Mendes, I. R., & Kasuya, M. C. (2015). Degradation of green polyethylene by Pleurotus ostreatus. *PLOS ONE*, *10*(6), e0126047. https://doi.org/10.1371/journal.pone.0126047.

Desai, J. D., & Banat, I. M. (1997). Microbial production of surfactants and their commercial potential. *Microbiology and Molecular Biology Reviews*, *61*(1), 47–64. https://doi.org/10.1128/mmbr.61.1.47-64.1997.

Dockendorff, C., Fuss, S., Agra, R., Guye, V., Herrera, D., & Kraxner, F. (2022). Committed to restoring tropical forests: An overview of Brazil's and Indonesia's restoration targets and policies. *Environmental Research Letters*, 17(9). https://doi.org/10.1088/1748-9326/ac8ab2.

Elis acirc ngela, M. R., Etiele, GdM., Cibele, FdO., Kellen, Z., & Jorge, A. V. C. (2015). Microalgae cultivation for biosurfactant production. *African Journal of Microbiology Research*, *9*(47), 2283–2289. https://doi.org/10.5897/AJMR2015.7634.

Fagbemi, O. K., & Sanusi, A. I. (2016). Effectiveness of augmented consortia of Bacillus coagulans, Citrobacter koseri and Serratia ficaria in the degradation of diesel polluted soil supplemented with pig dung. *African Journal of Microbiology Research*, *10*(39), 1637–1644. https://doi.org/10.5897/AJMR2016.8249.

Garg, S. K. & Tripathi, M. (2017). Microbial strategies for discoloration and detoxification of azo dyes from textile effluents. *Research Journal of Microbiology*, *12*(1), 1-19. https://doi.org/10.3923/jm.2017.1.19.

Gann, G. D., McDonald, T., Walder, B., Aronson, J., Nelson, C. R., Jonson, J., Hallett, J. G., Eisenberg, C., Guariguata, M. R., Liu, J. & Hua, F. (2019). International principles and standards for the practice of ecological restoration. *Restoration Ecology*, *27* (Suppl. 1) (Suppl. 1), S1–S46.

Gupta, R. K., Choudhary K. K., Kumar M., Negi A. & Rai H. (2012). Bioremediation & cyanobacteria: An overview. *Bionano Front.,* 9 :190–196. https://doi.org/10.13140/RG.2.1.2003.4323.

Hassan, M. M., Alam, M. Z. & Anwar, M. N. (2013). Biodegradation of Textile Azo Dyes by Bacteria Isolated from Dyeing Industry Effluent. *International Research Journal of Biological Sciences*, *2*(8), 27–31.

Hewson, I., Chow, C. & Furhman, J. A. (2010). Ecological Role of Viruses in Aquatic Ecosystems. *Encyclopedia of Life Sciences*. https://doi.org/10.1002/9780470015902.a0022546.

Jibran, A. K. & Milsee Mol J. P. (2011). *Pleurotus sajor-caju* Protein: A potential biosorptive agent. *Adv Bio Tech.,* 4 :25–27.

Keshani, Sharma, S., & Kaur, J. (2021). Microbial Diversity of Different Agroecosystems: Current Research and Future Challenges. *Microbiological Activity for Soil and Plant Health Management*, 65-89. https://doi.org/10.1007/978-981-16-2922-8_3.

Konishi, M., Fukuoka, T., Morita, T., Imura, T., & Kitamoto, D. (2008). Production of new types of sophorolipids by Candida batistae. *Journal of Oleo Science*, *57*(6), 359–369. https://doi.org/10.5650/jos.57.359.

Kristanti, R. A., Hadibarata, T., Toyama, T., Tanaka, Y., & Mori, K. (2011). Bioremediation of crude oil by white rot fungi Polyporus sp. S133. *Journal of Microbiology and Biotechnology*, *21*(9), 995–1000. https://doi.org/10.4014/jmb.1105.05047.

Kumar, A., Kaushal, S., Saraf, S. A., & Singh, J. S. (2017). Cyanobacterial biotechnology: An opportunity for sustainable industrial production. *Climate Change and Environmental Sustainability*, *5*(1), 97–110. https://doi.org/10.5958/2320-642X.2017.00011.4.

Luo, D., Xie, Y. F., Tan, Z. L., & Li, X. D. (2013). Removal of Cu2+ ions from aqueous solution by the abandoned mushroom compost of Flammulina velutipes. *Journal of environmental biology*, *34*(2 Spec No), 359–365.

Manchikanti, P. (2016). Nature conservation : Natural Lands and Biological Diversity. *Yearbook of International Environmental Law.* Convention on Biological Diversity (CBD), *27*, 219–225. https://doi.org/10.1093/yiel/yvx029.

McSweeney, C., & Mackie, R. (2012). Commission on genetic resources for food and agriculture. *Micro-organisms and ruminant digestion : State of knowledge, trends and futurs prospects. Background Study Paper (FAO)*, *61*, 1-62.

Miflin, B. J., & Lea, P. J. (1976). The pathway of nitrogen assimilation in plants. *Phytochemistry*, *15*(6), 873–885. https://doi.org/10.1016/S0031-9422(00)84362-9.

Mohamed, A. T., El Hussein, A. A., El Siddig, M. A., & Osman, A. G. (2011). Degradation of oxyfluorfen herbicide by soil microorganisms biodegradation of herbicides. *Biotechnology (Faisalabad)*, *10*(3), 274–279. https://doi.org/10.3923/biotech.2011.274.279.

Mohan, D., Abhishek, K., Sarswat, A., Patel, M., Singh, P., & Pittman, C. U. (2018). Biochar production and applications in soil fertility and carbon sequestration – A sustainable solution to crop-residue burning in India. *RSC Advances*, *8*(1), 508–520. https://doi.org/10.1039/C7RA10353K.

Mukherjee, S., Das, P., & Sen, R. (2006). Towards commercial production of microbial surfactants. *Trends in Biotechnology, 24*(11), 509–515. https://doi.org/10.1016/j.tibtech.2006.09.005.

Njoku, K., Yussuf, A., Akinola, M., Adesuyi, A., Jolaoso, A., & Adedokun, A. (2016). Mycoremediation of petroleum hydrocarbon polluted soil by Pleurotus pulmonarius. *Ethiopian Journal of Environmental Studies and Management, 9*(1), 865–875. https://doi.org/10.4314/ejesm.v9i1.6s.

Perelo, L. W. (2010). *In situ* and bioremediation of organic pollutants in aquatic sediments. *Journal of Hazardous Materials*, 177(1–3), 81–89. https://doi.org/10.1016/j.jhazmat.2009.12.090.

Prasad, Y. L. & Sachin, D. R. (2013). Biosorption of Cu, Zn, Fe, Cd, Pb and Ni by non-treated biomass of some edible mushrooms. *Asian Journal of Experimental Biological Sciences, 4*(2), 190–195.

Rai, Y., & Smith, D. (2016). Self-purifying properties of the Ganges River Scoping Review. *Western Undergraduate Research Journal: Health & Natural Sciences*, 7(1), 1-3.3.

Richardson, A. E., Barea, J.-M., McNeill, A. M., & Prigent-Combaret, C. (2009). Acquisition of phosphorus and nitrogen in the rhizosphere and plant growth promotion by microorganisms. *Plant and Soil, 321*(1–2), 305–339. https://doi.org/10.1007/s11104-009-9895-2.

Ron, E. Z., & Rosenberg, E. (2001). Natural roles of biosurfactants. *Environmental Microbiology, 3*(4), 229–236. https://doi.org/10.1046/j.1462-2920.2001.00190.x.

Safiyanu, I., Isah, A. A., Bubakar, U. S. & Rita Singh, M. (2015). Review on comparative study on bioremediation for oil spills using microbes. *Research Journal of Pharmaceutical, Biological and Chemical Sciences, 6*, 783–790.

Sayer, J., Sunderland, T., Ghazoul, J., Pfund, J. L., Sheil, D., Meijaard, E., Venter, M., Boedhihartono, A. K., Day, M., Garcia, C., van Oosten, C., & Buck, L. E. (2013). Ten principles for a landscape approach to reconciling agriculture, conservation, and other competing land uses. *Proceedings of the National Academy of Sciences of the United States of America, 110*(21), 8349–8356. https://doi.org/10.1073/pnas.1210595110.

Sharma, P. (2021). Efficiency of bacteria and bacterial assisted phytoremediation of heavy metals: An update. *Bioresource Technology, 328*, 124835. https://doi.org/10.1016/j.biortech.2021.124835.

Simarro, R., González, N., Bautista, L. F., & Molina, M. C. (2013). Assessment of the efficiency of *in situ* bioremediation techniques in a creosote polluted soil: Change in bacterial community. *Journal of Hazardous Materials, 262*, 158–167. https://doi.org/10.1016/j.jhazmat.2013.08.025.

Singh, A. K., Sisodia, A., Sisodia, V. & Padhi, M. (2019). *Role of microbes in restoration ecology and ecosystem services. New and future developments in microbial biotechnology and bioengineering* (pp. 57–68). Elsevier. https://doi.org/10.1016/B978-0-444-64191-5.00004-3.

Soffer, N., Woolston, J., Li, M., Das, C., & Sulakvelidze, A. (2017). Bacteriophage preparation lytic for Shigella significantly reduces Shigella sonnei contamination in various foods. *PLOS ONE*, *12*(3), e0175256. https://doi.org/10.1371/journal.pone.0175256.

Suding, K., Higgs, E., Palmer, M., Callicott, J. B., Anderson, C. B., Baker, M., Gutrich, J. J., Hondula, K. L., LaFevor, M. C., Larson, B. M., Randall, A., Ruhl, J. B., & Schwartz, K. Z. (2015). Committing to ecological restoration. *Science*, *348*(6235), 638–640. https://www.science.org/doi/10.1126/science.aaa4216.

Sulakvelidze, A. (2013). Using lytic bacteriophages to eliminate or significantly reduce contamination of food by foodborne bacterial pathogens. *Journal of the Science of Food and Agriculture*, *93*(13), 3137–3146. https://doi.org/10.1002/jsfa.6222.

Sutherland, C. & Venkobachar, C. (2010). A diffusion-chemisorption kinetic model for simulating biosorption using forest macro-fungus, fomes fasciatus. *International Research Journal of Plant Science*, *1*(4), 107-117. http://dx.doi.org/10.6084/m9.figshare.6197030.v1.

Thaniyavarn, J., Chianguthai, T., Sangvanich, P., Roongsawang, N., Washio, K., Morikawa, M., & Thaniyavarn, S. (2008). Production of sophorolipid biosurfactant by pichia anomala. *Bioscience, Biotechnology, and Biochemistry*, *72*(8), 2061–2068. https://doi.org/10.1271/bbb.80166.

Van Bogaert, I. N. A., Saerens, K., De Muynck, C., Develter, D., Soetaert, W., & Vandamme, E. J. (2007). Microbial production and application of sophorolipids. *Applied Microbiology and Biotechnology*, *76*(1), 23–34. https://doi.org/10.1007/s00253-007-0988-7.

Winterhalder, K., Clewell, A. F., & Aronson, J. (2004). Values and science in ecological restoration - A response to Davis and Slobodkin. *Restoration Ecology*, *12*(1), 4–7. https://doi.org/10.1111/j.1061-2971.2004.12001.x.

Chapter 6

Biotechnological Approaches in Mitigating Climate Variability and Anthropogenic Factors

Riddha Dey
Surya Prakash Dube
Seema Devi
and Richa Raghuwanshi*
Department of Botany, MMV, Banaras Hindu University, Varanasi, Uttar Pradesh, India

Abstract

Rapid climate change and harsh anthropogenic activities have affected the ecosystem beyond resilience. Human-natural systems interference has brought about a large number of negative impacts across the continents, which includes the rise in the sea level, biodiversity loss, altered ecosystem; structure and function, and human livelihoods. Many species have got extinct and many are listed as endangered. To protect them and restore the balance of the environment, it is crucial to sensitize humans and encourage the applications of biotechnological tools in ecosystem restoration. The interference of biotechnology in agriculture has helped in developing climate-resilient crops. The techniques like genetic engineering, meta-omics, and CRISPR Cas9 are some of the promising conservation technologies that are commercially affordable, reproducible, and need to be implemented at the pilot level. This chapter may help in representing possible approaches to improve adaptive capacity as well as restoration of the ecosystem. It deals with various effects of anthropogenic activities and climate changes on different ecosystems like the Himalayan ecosystem, desert ecosystem,

* Corresponding Author's Email: richabhu@yahoo.co.in.

In: Conserving, Halting and Reinstating Degraded Natural Resources …
Editors: Rajani Srivastava and Pankaj Kumar
ISBN: 979-8-89113-067-8

terrestrial ecosystem, and aquatic ecosystem. It also focuses on the biotechnological techniques which can be used to diminish biodiversity loss and help restore the ecological balance.

Keywords: anthropogenic, bioremediation, biotechnology, biodiversity, climate change, ecosystem restoration

1. Introduction

Evaluation of future climate change brought by human activities is based on projections of how the world's energy system will change during the coming century. Future energy production is fundamentally impacted by anthropogenic, greenhouse gas (GHG) emissions and global warming. Plasticity and adaptations may help the biotic systems to establish and expand their populations in new geographical areas. Increased temporal climate change might benefit genotypes with adaptive plasticity (Anderson et al., 2012). Climate change has impacted the biogeochemical processes as well as microbial diversity. Anthropogenic pollution due to urbanization and industrialization has changed ecological health decreasing the fitness of flora and fauna, loss in genetic variation and gradually taking toward extinction. Climate change adds to the adversities of an ecosystem already bearing stress from human usage which can jeopardize the long-term success of restoration initiatives. Management goals and adaptation plans are now progressively taking climate change's effects on populations and communities into account (Tong et al., 2014). Restoration strategies that emphasize resilience which is well understood as resistance to change and recovery from change, can promote climatic adaptation. Enhancing restoration and monitoring efforts may be possible by using resilience concepts and traits. The Special Report on Emission Scenarios (SRES) was published by the United Nations' Intergovernmental Panel on Climate Change (IPCC) in 2000. The Special Report on Emission Scenarios is still a crucial component of climate change modeling, as the most recent IPCC reports have shown (IPCC, 2001; 2007).

Biotechnological methods are a promising substitute for the decontamination process of the environment. For the degradation of recalcitrant and xenobiotic pollutants, the overlooked microbial genetic resources can be explored and exploited. Recent developments in "designer biocatalysts" for environmental restoration can be achieved by the molecular genetics of biodegradation and approaches to protein modification. However, due to the dangers associated with uncontrolled growth and proliferation of the added biocatalyst and horizontal gene transfer, the use of such genetically engineered microorganisms (GEMs) in the environment has been restricted. However, designing the programed death of biocatalysts, soon after the pollutant is depleted will reduce the associated risks, considering these technologies for effective bioremediation. Many conservation technologies are also used for ecological restoration. Gene modification, gene editing tools, and the OMICS approach are also effective technologies in manufacturing climate-resilient crops, which are briefly described in this chapter.

2. Climate Change and Anthropogenic Effects on Major Ecosystems

Climate includes the average temperature, amount of precipitation, days of sunlight, and other variables that might be measured at any given site. Climate changes include the changes in weather patterns, which associate directly or indirectly with human activity that alters the composition of the global atmosphere. Human demand for food and other livelihood products has taken up an enormous share of the earth's land area. The massive conversion of forests, wetlands, grasslands, and other terrestrial ecosystems has accelerated the extinction rate. It is speculated that the increase in annual mean temperatures will be the biggest challenge in plants, animals, and human survival, soon. The effects on major ecosystems are discussed below.

2.1. Effects on the Himalayan Ecosystem

The Himalayas, which literally means the abode of snow, is the youngest and the highest range of Fold Mountains in the world. A great diversity of flora and fauna is found in Indian Himalayas comprising about 45,000 species of plants and 65,000 species of animals found in India. Therefore, this great strength of flora and fauna put the country on the list of mega biodiversity centers of the world. The flora and fauna of the Himalayas vary with variations in climate.

The most widely reported adverse effect of climate change on the Himalayan ecosystem is the rapid reduction of glaciers. Other ongoing anthropogenic disturbances reported are fire (Gupta, 1978), cutting (Awasthi et al., 2003), overgrazing (Nautiyal et al., 2004), deforestation (Kumar & Ram, 2005), intensive agriculture (Semwal et al., 2004), shifting-cultivation (Zonunsanga et al., 2014), and other land-use practices (Figure 1).

The ever-increasing demand of the population for resources has put pressure on the biological resources of the world. The major causes of species extinction are habitat loss and fragmentation. *Catreus wallich*, of the Jammu and Kashmir regions, is declining due to the loss of habitat (Wikramanayake et al., 1998). The genus *Schizothorax* contains at least six endemic species in high mountain lakes and streams. The genus *Ptychobarbus* and the Ladakh snow trout (*Gymnocypris biswasi*) are now thought to be extinct. Change in climate has also increased a large number of infectious diseases transmitted by insects, due to habitat shift. Agriculture, horticulture, and age-old traditional food crops are also destroyed (Eriksson, 2008). An increase in atmospheric temperature accelerates the growth of crops and shortens the growth period. In cereal crops, such changes have led to reduced yield. Climate change in the Himalayan region has increased the extinction rate of flora and fauna and changed the rainfall pattern, duration of vegetative growth, maturity period, and overall growth of crop plants. A 2019 report by Intergovernmental Sciences-Policy Platforms as Biodiversity and Ecosystem Services noted that due to human activities, one million plant and animal species are facing extinction.

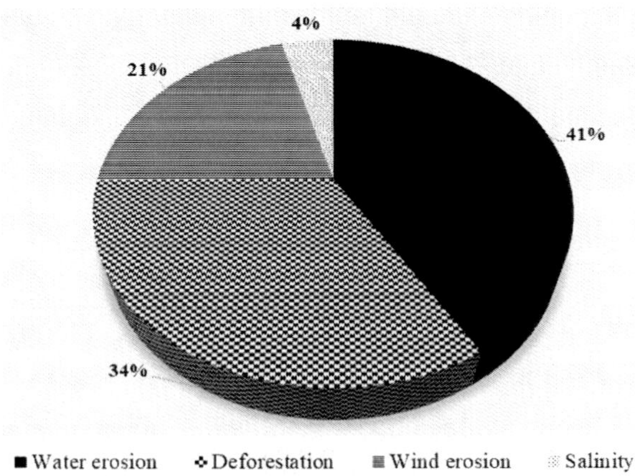

Figure 1. Percentage of the area affected due to climate change and anthropogenic activities in the Himalayan ecosystem.

Continuous warming is already occurring at approximately three times the global average (Nogues-Bravo et al., 2007). The Climate report of the United States National Oceanic and Atmospheric Administration (NOAA, 2009), shows some evidence about the warming of the earth progressively. The evidence is based on 10 indicators of which seven are increasing (air temperature in the troposphere; specific humidity; Ocean heat content; rising sea level; sea-surface temperature; temperature over oceans; temperature over land) while rest three indicators are decreasing (extent of snow cover; volume of glaciers and sea ice). In the Himalayan region, the temperature increase at higher altitudes is greater than in the lowlands (Shrestha et al., 1999). The recent Fourth Assessment Report of the Intergovernmental Panel on Climate Change (IPCC) shows a 0.74°C rise in temperature over the past 100 years (1906-2005).

Rising global temperature has resulted in a large number of changes in the energy exchange processes (radiative forcings), affecting atmospheric circulation and global precipitation patterns (Beniston, 2003). In the northern hemisphere over mid and high latitudes, there is an overall increase of 5-10% in annual precipitation and a decrease by an average of 3% in subtropical areas (IPCC, 2001; Vinnikov et al., 1990). In the north-western Indian Himalayas during the period 1866-2006, Bhutiyani et al., (2010) reported a statistically significant downward trend in monsoon with average annual rainfall in the north-western Indian Himalayas during the period 1866-2006. A similar trend was observed for the period 1960-2006 in the western Indian Himalayan region (Sontakke et al., 2010). Dimri & Dash (2012) observed a significant decrease in winter precipitations between December and February in the region during the period 1975 - 2006 reporting an increasing number of warm days and a decreasing number of cold days in the western Indian Himalayas. An increase in pre-monsoon (March-May) was also observed in the western Indian Himalayas during 1901-2003 (Guhathakurta & Rajeevan, 2008).

Melting of glaciers in the Himalayas contributes to the rise in sea level and an increase in the frequency of glacier hazards such as glacial lake outburst floods, avalanches, landslides, and permafrost creep. Of the greater Himalayan region, only 17% is now covered by glaciers and is receding more rapidly (Dyurgerov & Meier, 2005; IPCC, 2007). If exerting increase in temperature continues, glaciers located on the Tibetan Plateau are likely to shrink from 500,000 km^2 (the 1995 baseline) to 100,000 km^2 or less by the year 2035. Increased melting results in increased water flow. With time, the glaciers may completely disappear or approach a new equilibrium, bringing long-term effects like increasing water shortages and limited supplies for downstream communities, particularly during the dry season. The rivers most likely to experience the greatest loss in water availability due to melting glaciers are the Indus, Tarim, Yangtze, Brahmaputra, and Amu Dary.

2.2. Effects on Desert Ecosystem

Both global climate change and anthropogenic activities are the main driving forces for desert ecosystems. Deserts are destroyed by a variety of factors including livestock grazing, the introduction of exotic species, off-road vehicle use, urbanization, and its attendant effects.

From a very past time, deserts were destroyed by livestock and grazing. The impacts of grazing are very extensive whether it may be positive or negative. According to a recent biological investigation in the western Mojave Desert of California, 100% of a 234-square-kilometers area was affected to some extent by sheep grazing (Tierra Madre Consultants, 1991). Response of plants to grazing varies according to species, season, plant phenology, local conditions (drought, edaphic factors, etc.), and past historical use (Genin & Badan-Dangon, 1991). The direct effects of grazing on desert animals are not well documented. Grazing sheep can damage tortoise burrows. Grazing also increases the spatial and temporal heterogeneity of water, nitrogen, and other soil resources, fostering increased desertification of productive arid lands.

Constructions of roads or highways, railway tracks, powerlines, and pipeline transport are some of the most conspicuous elements of desertification. Some other negative effects of corridors include direct mortality of animals by hitting vehicles on roadways, habitat fragmentation which restricts the movement of animals and flow of genes between populations, and increased access to remote areas for illegal collection or vandalism of plants and animals (Boarman & Sazaki, 1996), and increased soil erosion (Wilshire & Prose, 1987). Corridors can also serve as a source of exotic invasive plants brought in or on construction equipment (Zink et al., 1995). Invasive plants prosper in disturbed conditions and contribute to an increased likelihood of fire. The construction of pipelines for gas, oil, and water is much more destructive than overhead lines because extensive trenching is usually required. This traditionally has led to severe soil impacts (leaving subsoil on the surface), disturbing stabilized crusts and rock surfaces, and concentrating runoff and erosion. Another effect of roads is edge enhancement in which perennial shrubs along roadsides are denser, larger, more vigorous, and support greater numbers of foliage arthropods than those away from roadsides (Lightfoot & Whitford, 1991).

Sometimes a large area of the desert ecosystem is destroyed by ongoing military activities. A study of the California desert showed that between the time periods of 1942-1944, a large number of soldiers passed through these training facilities, which covered more than 46,800 square kilometers (Bureau of Land Management 1990). All impacted areas exhibited significant reductions in plant density and cover relative to control areas. Long-lived species such as Larrea tridentata were dominant in all control areas but percentage cover and density were reduced in impacted areas. Dominant plants in disturbed areas included pioneer species such as *Ambrosia dumosa* and *Hymenoclea salsola.* Differences in vegetative structure between control and impacted plots were due to soil compaction, changes in soil texture, removal of the topsoil layer, and alteration of drainage channel density (Prose et al., 1987).

Invasive exotic plants have shown a great contribution toward the destruction of the southern California desert ecosystem (Rejmánek & Randall, 1994). Invasion of exotic species results in habitat disturbance. Once established, the exotic species may decrease the abundance of native species due to competition. Some important exotic plants in southern California are salt cedar (*Tamarix ramosissima*), also known as tamarisk (Lovich & de Gouvenain, 1998), Russian thistle (*Salsola iberica*) (Young, 1991), filaree (*Erodium cicutarium*), and also a large number of 42 grass species including split grass (*Schismus* spp.) and bromes (*Bromus* spp.) (Hunter,1991). Exotic plants show two major problems in the desert ecosystem. First, they increase the fuel load and frequency of fire in a community that is poorly adapted to fire. Second, some exotic plants also show allelopathic effects that negatively affect native species.

2.3. Effects on Terrestrial Ecosystem

Climate change affects the pattern and dynamics of energy flow and material circulation in several ecosystems including terrestrial ecosystems. Some anthropogenic activities such as overgrazing, urbanization, mining, land-use changes, and deforestation have greatly impacted the terrestrial ecosystem productivity. Climate change activities such as rising temperatures and altered rainfall patterns can cause large-scale forest dieback. The long-term drought (40-50% reduction in annual rainfall), an acute drying and warming, and a series of heat waves in the summer of 2010/11 resulted in significant canopy dieback across a range of forest and woodland types, including *Eucalyptus marginata*, *Corymbia calophylla* (Matusick et al., 2013), *Banksia attenuata*, *Banksia menziesii* (Bader et al., 2014; Challis et al., 2016) and *Eucalyptus gomphocephala* (Matusick et al., 2013).

Vegetation of the alpine ecosystem is at high risk from climate change. The Bogong High Plains in Victoria having the longest-running monitoring plots in Australia due to environmental changes showed a shift in vegetation. Grasses and other graminoids got decreased by up to 25% between 2004 and 2010. Forb and shrub cover increased by 9 and 20%, respectively (Wahren et al., 2013). This sharp decline in grass cover was mostly credited to severe drought, with increases in shrubs and forbs credited to increases in temperature (Camac et al., 2015). In addition to changes in vegetation composition, warming altered Alpine flowering phenology (Hoffmann et al., 2010).

The seasonal timing of growth or reproduction of plant species is often influenced by temperature and is therefore influenced by climate change. Climate change is also responsible for phenological changes in some plants. Orchids are known as climate change indicators. Orchid phenology is particularly sensitive to seasonal temperature and day length and their phenological shifts are analyzed through the assessment of herbarium specimens (Robbirt et al., 2011). One main threatening process for wild orchid species is poor reproductive success (Faast & Facelli, 2007), caused by factors that are likely exacerbated by climate change. Empirically detected phenological trends in orchids such as decoupling of orchid–pollinator interactions, increased hybridization, and disruption of whole community networks are also caused by climate change.

The impact of industrialization on the terrestrial ecosystem is well known. Industrialization is one of the main factors contributing to ecosystem destabilization and therefore needs special attention. Industrial effluents include a large number of pollutants such as heavy metals and microplastics. A single production plant in the Austrian portion of the Danube River could emit 94.5 t/year of industrial microplastics considering an effluent production of 100 L/s (Lechner & Ramler, 2015). The untreated domestic sewage is rich in fibers from clothing and microplastic beads from personal care products, among others (Mason et al., 2016). The sludge also contains 80-90% of microplastics. On interacting with soil fauna microplastics also change their biophysical environment, with potential consequences for their fitness and soil function (Zhu et al., 2018). Springtails and earthworms have been reported to transport microplastics within the soil in both horizontal and vertical directions (de Souza Machado et al., 2018). For earthworms, microplastic exposure was associated with structural changes in their burrows and also shows a reduction in their growths (Huerta Lwanga et al., 2017). Changes in the biophysical environment also show an adverse effect on the gut microbiomes of springtails (Zhu et al., 2018). Therefore, even without clear evidence of ingestion, microplastic-exposed springtails had altered gut microflora. Most plastic materials leach compounds such as phthalates and bisphenol A show estrogenic activity and also show the activity of endocrine disruption in vertebrates and some invertebrate species (Sohoni & Sumpter, 1998).

The increased and/or day-by-day increasing concentration of greenhouse gasses such as carbon dioxide, water vapors, nitrous oxide, and methane, trap more solar radiation in the atmosphere, reflecting from the earth's surface and sending it back to the earth's surface resulting in the rise of average global temperature is referred to as global warming (Mukhopadhyay, 2013). The burning of fossil fuels, deforestation, and volcanic activities are the main reasons for global warming. Global warming is the main cause of the melting of ice in the Polar Regions resulting in the rise of the ocean level. The political response to global warming is varying. A large number of international frameworks have been developed for the mitigation purpose of global warming, such as the 2016 Paris Agreement seeking to limit temperature increase. According to the Australian Greenhouse Office, 80% of the world's total energy needs are met by fossil fuels, namely oil, coal, and natural gas. On the other hand, human activities contribute 100 times more to the release of carbon dioxide than volcanoes. Intergovernmental Panel on Climate Change (IPCC) study, shows a 1°C rise in global temperature for the emission of 500 Gigaton (Gt) of carbon dioxide into the atmosphere by 2020 (IPCC AR5, 2013). The release of more than 1000 Gt of carbon dioxide generated by fossil fuels may show a 2°C rise in temperature by 2050 (IPCC AR5, 2013). The phenomenon of temperature rise contributes to the melting of polar ice and rising global sea level. These frequent droughts, floods, and tropical cyclones result in major hazards to human health, the environment, water resources, agriculture, and infrastructure (hospitals, industries, and educational institutions).

2.4. Xenobiotics

Xenobiotic (Xenos: foreign or strange and bios: life) is a term used to describe chemical substances that are foreign to animal life. Major Xenobiotics contaminants received from industries are halogenated and nitrated hydrocarbons. The unusual presence of any substances in large quantities is also considered in the xenobiotic category (Sikandar et al., 2013). Xenobiotics may remain conserved for a long duration (months to years) in natural environments in the form of polymers or spores even in the aquatic ecosystem as pollutants. Some of the xenobiotics regularly affecting biotic communities are pesticides, phenols, plastics, hydrocarbons, fuels, and polyaromatic compounds (Bulucea et al., 2012; Serra et al., 2013). Detectable levels of xenobiotic compounds responsible for more than 13 million deaths and 24% of world diseases every year are either present in edible substituents or water (Banjoko, 2014). Xenobiotic compounds especially pharmaceuticals which are non-decomposable during sewage treatment adversely affect all water storage bodies. There are several non-steroidal anti-inflammatory drugs (NSAIDs) such as naproxen, ibuprofen, diclofenac, acetylsalicylic acid, and ketoprofen which have altered gene expression (epigenetic) or organ histology of the aquatic biotic community (Cuklev et al., 2012; Shanmugan et al., 2013). DDE (dichloro diphenyl dichloroethylene: induced eggshell thinning in birds) and DDT complex (dichlorodiphenyltrichloroethane: ovo-testis in male western gulls) are referred to as most concerning EDCs (endocrine disruptors) (Le Page et al., 2011).

2.5. Effects on Aquatic Ecosystem

The division of pollution into categories like air, water, and land was deeply criticized and it was defined as "one pollution" as every pollutant ends up in the ocean no matter where it originated (Williams, 1996). Pollutants usually emerge from industries, construction sites, and human settlements, which flow into the water bodies. The main constituents of pollutants are heavy metals, organic pollutants, sediments, oils, radionuclides, etc. (Williams, 1996). These contaminants are toxic and harmful jeopardizing the organisms and environment of the global ocean resulting in high-scale disturbances in the aquatic ecosystem.

Due to climate change, the distribution and abundance of aquatic ecosystems has got disturbed. Climate change causes an increase in thermal stratification, and a rise in sea level resulting in ocean warming. The direction of ocean water movement is strongly influenced by the density of seawater, which is altered by the changes in global warming. The organisms in the marine ecosystem fail to acclimatize themselves after their optimal temperature, beyond which their fitness is reduced, the mortality rate increases, and they are gradually driven towards extinction risk (Hochachka & Somero 2002). It has been reported that phytoplankton distribution and abundance were deeply affected by the temperature rise (Ducklow et al., 2009). The annual productivity of the World Ocean showed a 70% decline, specifically in the Pacific and Indian Ocean (Polovina et al., 2008). A rise in temperature also strengthens the consumer control of primary production, as a result, total food web biomass is drastically reduced (O'Connor et al., 2009). In Heron Island, the coral reef ecosystem is declining due to an abrupt increase in sea temperatures. As a result, there is an increase in coral bleaching and mortality, which in turn gives rise to ocean acidification. In another study, it was found the rising ocean temperature deeply affected the distribution of Kelp Forest near San Clemente Island, California (Hoegh-Guldberg & Bruno, 2010).

Aquatic ecosystem forms complex habitat-forming tendencies. The species like corals, sea grass, oysters, mangroves, and marsh grasses form complex habitats together to maintain aquatic ecosystems. The disappearance of any of the organisms in these habitats can lead to disturbances to the whole ecosystem. For example, a rise in ocean temperature leads to mass coral bleaching, which is leading to the disappearance of coral reef fishes (Hoegh-Guldberg et al., 2007). Coastal angiosperms including mangroves, sea grass, and salt-marsh communities are exposed to threats due to climate changes. Deforestation of mangroves also leads to a reduction in wetland habitats. Climate change affected the timing of phytoplankton blooms, reducing phytoplankton and sea alga habitat in Polar Regions, which disturbed the polar marine food web dynamics (Gradinger, 2009). Sea ice plays an important role in providing a habitat for a variety of birds, mammals, and migratory animals. It also contains coral reefs and kelp forests. But a loss in sea ice due to global warming resulted in a prominent decrease in the polar biodiversity. Antarctic warming resulted in the declining number of penguins, seals, as well as polar bear counts reduced to 68% from their summer habitat (Barbraud et al., 2008).

The freshwater ecosystem also faces biodiversity loss and disruption in the natural habitat of aquatic organisms. Warming affects the stream habitat including the cold-water fishes like trout, charr, and salmons. This also affects their eggs and larvae due to a decrease in the oxygen level of water. Construction of dams and illegal water obstruction resulted in reduced flows, weed growth, wetland drainage, river fragmentation, and degradation of habitat and freshwater biodiversity.

Wetland ecosystems are very susceptible to climate change. Increase in temperature and a decrease in precipitation result in wetland loss (Liu et al., 2014). Global warming, results in decreased precipitation leading to the degradation of wetlands and a depleted environment. In China, a study was conducted on Sanjiang Plain using remote sensing by Mao et al., (2021). Sanjiang National Nature Reserve and Honghe National Nature Reserve are two internationally important reserves. Sanjiang Plain was once the largest freshwater wetland, but now a severe depletion is recorded. During the period from the 1950s to the 2000s, the Sanjiang Plain experienced severe land use and large-scale agricultural development which is responsible for the 77% loss of wetlands and fragmentation of wetland landscapes (Wang et al., 2011). In addition, when eight farms were built, 7317 km^2 of wetlands were lost in Sanjiang Plain.

In India, Srinagar's Dal Lake and Loktak Lake of Manipur have shown degraded quality of water due to eutrophication and weed infestation which has depleted the ecological functioning and biodiversity of these lakes. Water hyacinth is a dangerous invasive species that infested about 75% of the wetland area in Harike Lake (Punjab). Unsustainable exploitation of natural resources with activities like excessive fishing creates a disturbance in the wetland ecosystems. Due to resource encroachment and unplanned land use practices, Loktak Lake in Manipur valley is dangerously threatened. Above all, the maximum area of the lake is choked with various invasive species including *Eichhornia crassipes*, *Typha angustata*, *Hydrilla verticillata*, and *Chara* sp. Also, in Kaliveli, Tamil Nadu, due to land encroachment for harvesting paddy and illegal poaching practices for meat has led to a reduction of the water-spread area and the number of water-bird species like pelicans, storks, and flamingos as well it increases their migration. A large number of lakes including Kolleru Lake of Andhra Pradesh, Deepar Beel of Assam, Pyagpur and Sitadwar Jheels of Uttar Pradesh, and Hokarsar Lake of Kashmir have shrunk due to illegitimate use of land for agriculture.

3. Tools for Ecological Restoration

3.1. Transgenic Plants

Organisms undergoing genetic manipulations through genetic engineering techniques are called transgenic organisms. A foreign DNA containing the gene of interest is introduced inside the plant cells and inserted into the genome. These can either up-regulate or down-regulate the existing activity of the cell. Promoters help to regulate the expression of the introduced gene. Moreover, the marker gene is also inserted in the gene construct for the successful selection of transgenics post-transformation. These marker genes usually possess antibiotic or herbicidal properties. The introduced foreign DNA integrates into the host genome and is inherited by the next progeny. In laboratories and greenhouse experiments, *Arabidopsis thaliana*, tobacco (*Nicotiana tabacum*), or Indian mustard (*Brassica juncea*) are used as model plants for testing transgenic approaches.

Due to the harmful effects of anthropogenic activities and climate change, plants undergo various abiotic stresses. The plant cells accumulate certain low molecular mass organic compounds in cytoplasm known as "compatible solutes" in order to acclimate themselves under stressed conditions. An amphoteric quaternary amine, glycine betaine (GB) is an important compatible solute in plants. It has been reported that certain plants lack the production of GB which makes it difficult for them to escape from the damaging effects of stress. Genetic engineering made it possible to introduce the GB biosynthetic pathway in GB-deficient plants in order to provide them with stress protection (Sakamoto & Murata, 2002) and help in survival under changing climatic conditions.

Genetically manipulated plants, in which the trinitrotoluene (TNT) explosive degradation gene from bacteria was engineered and expressed, were found to tolerate and degrade TNT (French et al., 1999). Trichloroethylene, a chlorinated solvent, is a pronounced groundwater and soil contaminant. The transgenic plants which were engineered with mammalian cytochrome P450, resulted in improved trichloroethylene metabolism, helping in terrestrial ecosystem restoration (Doty et al., 2000). It has been reported that the overexpression of membrane transporter protein and chelator molecule overproduction aided the transgenic plants to tolerate and accumulate inorganic pollutants (Dhankher et al., 2002; Hirschi et al., 2000; Song et al., 2003). Genetic engineering is the key process to augment plant metabolite volatilization. Overexpression of cystathionine-gamma-synthase and selenocysteine methyl transferase in transgenic plants has led to the conversion of selenocysteine into volatile selenium. The introduction of the Mer A gene (mercury reductase) into the plant genome resulted in visible mercury tolerance and volatilization of elemental mercury (Rugh et al., 1996). It is also necessary to engineer plants with fast-growing and high biomass-producing capabilities to tolerate pollutants. Engineered plants have improved root shoot translocation capacity which helps in pollutant remediation.

3.2. Climate Resilient Crops

The genetic engineering approach has helped to develop crops with climate-resilient potential enabling them to tolerate the adverse effects of climate change and aid in ecological restoration. The modified crops develop pathways to tolerate biotic/abiotic stresses as well as to resist the effects of greenhouse gases, drought, flood, pesticide, and fertilizer (Barrows et al., 2014), addressing the issues of food security under changing climatic conditions.

Submergence tolerant rice was developed by IRRI, by engineering the SUB 1A gene into it, which allowed the rice plant to withstand water for up to 2 weeks, still producing a higher yield (Stone, 2008) thereby giving a solution to rice cultivation in flooded soils. Drought-resistant maize, "MON 87460" or "Drought Guard" developed by Monsanto, is capable of tolerating robust drought in the United States and maintaining productivity. A number of successful projects were worked out including "Drought Tolerant Maize for Africa" (DTMA), "Water Efficient Maize in Africa" (WEMA), and "Improved Maize for African Soils" (IMAS) (Varshney et al., 2011). These varieties showed a 20-50% high yield under drought stress.

A salt-tolerant transgenic maize line was produced by overexpressing 40 transcription factors. They showed improved photosynthesis, regulated stomatal conductance, and high chlorophyll content along with increased yield (Nelson et al., 2007). An aquaporin coding gene NtAQP1 is present in Tobacco (*N. tabacum* L.), which showed salt tolerance. In genetically modified tomatoes, the presence of this gene exhibited protection against salinity stress, and shoot/root hydraulic failure prevention (Sinclair et al., 2004). GM crops also help to sequester atmospheric carbon dioxide. An herbicide-resistant Roundup Ready™ soybean crop proved to sequester about 63,859 million Tonns carbon dioxide in the United States and Argentina (Brimner et al., 2005).

3.3. Population Genomics

For ecological restoration, it is important to know the indigenous location of plants. Due to the erosion of genetic diversity and increased inbreeding, native plant populations have got fragmented. Moreover, due to continuous climate change, the local provenance of plants has also got altered. As a result, there is a shift in the entire biome which has led to the mixing of provenances. So, conserving the species gene pool is the time demand. A better understanding of the geographical extent of local provenances, the limit of transfer to minimize maladaptation, and spatial patterns of adaptive variation will aid in successful plantings and restoration.

Population genomics significantly improves seed sourcing. The experimental field trial is a conceptually straight forward process to facilitate adaptation and understand evolutionary processes. It has been reported that experimental plantings are used to develop species transfer functions within the provenances to escape maladaptation (Wang et al., 2010; Yang et al., 2015). Fine-scale historical demographic information between associating population structure is very useful for improving resolution to delineate the origin (Rossetto et al., 2019; Supple et al., 2018). Population genomics also aids to identify genetic adaptations that increase the potential of plants to adapt through environmental changes (Breed et al., 2019). Some of the significant ways for ecological restoration by understanding the population genomics are enlisted in Table 1.

3.4. Seed Banks

Seed banks are a significant tool for plant conservation. Climate change has resulted in massive range shifts, habitat fragmentation, and pollinator loss which have made it difficult to maintain natural plant populations. Seed banks are ex-situ conservation of seeds that maintains intact plant populations as well as preserve the gene flow within species. Continuous collection and sample storage of all the local species will ensure genetic conservation and ecological restoration. Svalbard "doomsday" Seed Vault is one of the rarest seed banks constituting 4000 plant species, which is developed deep under the permafrost near the North Pole. The United States has a Native Plant Germplasm System, constituting 20 storage facilities, which store 15,000 agronomically important species. Millennium Seed Bank comprising 13% of the world's wild plant species is also contributing to biodiversity conservation.

3.5. Case Studies

The central north zone of Chile is one of the global biodiversity hotspots which include the temperate rain forest of southern South America and the Juan Fernandez Archipelago (Arroyo et al., 2004). About 19% of the biodiversity is reserved through National Parks and Natural Reserves but the maximum ecosystem is not protected through these in-situ conservation systems. Mexico is ranked fifth among the countries with rich plant diversity at the global level, but it is also ranked sixth in endemism rate, with 40% of its flora restricted within the country and is on the verge of extinction. Here are some examples of the strategies used by seed banks for ecological restoration in Latin America and the Caribbean region.

The Convention on Biological Diversity in Chile was implemented through the National Strategy for Biological Diversity (CONAMA 2003). The main objective was to protect and preserve species, and genetic diversity and promote sustainable resources use through ex-situ conservation. Instituto de Investigaciones Agropecuarias (INIA gene bank) started a conservation program for plant genetic resources in 1990 which is focused on protecting genetic diversity, through which it has preserved 54,000 germplasms of major economical crops including cereals and legumes (Salazar et al., 2006).

In 2001, a seed conservation program was developed in collaboration with INIA and Royal Botanical Garden, Kew with the global MSB partnership. The purpose of this collaboration was to conserve the diversity of endangered plant species on the basis of their endemism and phylogenetic uniqueness (Guerrero et al., 2008). The medicinal and ornamental plants as well as the plant species which were needed for restoration of degraded habitats were given priority in this conservation program.

Table 1. Ecological disturbances and mechanisms for restoration

Ecological disturbances	Target species	Restoration mechanism	Findings	Reference
Inbreeding	Bighorn sheep (Ovis canadensis)	Bottleneck tests	Avoiding inbreeding can decrease the chances of a declining population.	Ramey et al., (2000)
Endangered species	Seagrass (Zostera noltii)	Assignment tests using microsatellite markers	Successful donor populations are chosen for restoration based on shoot or seed transplantation. This will improve the evolutionary potential of an endangered species.	Diekmann et al., (2010)
Maladaptation	Common reed (Phragmites australis)	Regression of allele occurrence and Environmental Variables	Identification of genetic sources for ecological restoration.	Gao et al., (2012)
Invasive species	Brown rat (Rattus norvegicus)	Migration rates and assignment tests	Eradication is feasible with a low risk of recolonization.	Robertson & Gemmell (2004)
Catastrophic disturbances	Coastal tailed frog (Ascaphus truei)	Regression of least-cost paths and genetic differentiation	Natural regeneration maintains genetic diversity better than active management.	Spear et al., (2012)
Immigration	Australian rats (Rattus fuscipes and Rattus leucopus)	Assignment and clustering tests	Ecological corridor helps in habitat restoration and monitoring corridor efficiency to re-establish gene flow.	Paetkau et al., (2009)
Habitat degradation	Terrestrial orchid (Dactylorhiza incarnata)	Assignment and clustering tests and genetic parameters	Colonization patterns decrease genetic diversity but not population fitness.	Vandepitte et al., (2012)

Management of invasive species and plant biodiversity conservation is the main objective of Mexican National Strategies. CONABIO 2000, the National Strategy program, is based on inventories, workshops, and data collection. Total 22 groups of actions were involved in this seed conservation process. The National Strategy for Invasive Species focuses on controlling and eradicating invasive species. The Mexican National Strategy for Plant Conservation supports the initiation of GSPC (Secretariat of the Convention on Biological Diversity, 2010). Under this strategy, National University of Mexico (UNAM) established the Seed Bank “FESI-UNAM” (Facultad de Estudios Superiores Iztacala, UNAM), which particularly focuses on the storage and conservation of economically important wild plant species in endemic areas. It also generates and compiles morphological and physiological seed data to maintain an enriched seed bank. The main objective also includes the long-term preservation of seeds. This also promotes a continuous capacity-building program which increases the scope of research to improve the seed bank.

3.6. Marker-Assisted Genome Selection

The genetic methods for ecosystem restoration require reliable and accurate genome information. Godefroid et al., (2011) reported the importance of genetic information in improving restoration outcomes. Molecular markers are considered efficient tools for the identification of polymorphisms among different genotypes (Jiang et al., 2015). These have gradually become the fundamental tool for crop improvement. These can be categorized into two main types, non-PCR based (to identify RFLPs) and PCR based. PCR-based includes amplified fragment length polymorphism (AFLP) markers (Bandelj et al., 2004), single-nucleotide polymorphism (SNP) markers (Jin et al., 2003), intron length polymorphic (ILP) markers (Zhang et al., 2017), random amplified polymorphic DNA (RAPD) markers, and simple sequence repeat (SSR) markers (Wu et al., 2016; Min et al., 2017). SSR markers (microsatellites) are predominantly used in genetic applications, like map construction, fingerprinting, and genetic diversity analyses, due to their high reproducibility, polymorphism, abundance as well as their codominant nature. These are widely used for genetic diversity analysis, germplasm identification, comparative genetics analysis, phylogenetic analysis, QTL analysis, linkage mapping, and marker-assisted selection (Liu et al., 2010;

Mondal et al., 2015). These help to broadly study genome mapping, fingerprinting, population genetics studies, and molecular breeding.

MicroRNAs (miRNAs) a class of noncoding RNA, 18-24 nucleotides in length are responsible for regulating gene expression (Neilson & Sharp, 2008). Endogenous miRNAs being highly conserved sequences play essential roles in plant development and gene expression under diverse biotic and abiotic stress conditions (Khraiwesh et al., 2012; Singh et al., 2017). *Cleistogenes songorica* of *Gramineae* family was used to study the significance of SSRs in the miRNA gene. The study of miRNA-SSRs has given a brief idea about the cold and drought resistance in *C. songorica* (Zhang et al., 2014). *C. songorica* has the potential to play a critical role in landscaping, vegetation, and desert ecosystem restoration.

3.7. Meta-Omics for Monitoring Functional and Cryptic Ecosystem Components

Ecological restoration projects are justified in the view that they return both native biodiversity as well as ecosystem functioning which encompasses plant-microorganism, plant–insect symbioses, nutrient cycling and soil and substrate formation, and so on. The restored biodiversity and ecosystem functions must have the potential to provide ecosystem services, such as crop pollination, nutrient availability, soil stabilization, water purification, and disease regulation (Suding, 2015). To ensure the ongoing success of restoration efforts the biological components of ecosystems that provide these services should be the targets of restoration monitoring and assessment (Collen & Nicholson, 2014). Using field-based restoration monitoring and assessment approaches many biotic components of ecosystems (such as bacterial, fungal, and insect communities) are either difficult or impossible to survey. Expert taxonomists are required for field-based methods, such as sweep netting and/or blue-vane traps for flying insects and so on. Under these circumstances, genomic and meta-omic approaches have enabled taxonomic expert groups to accurately and rapidly characterize and quantify the bulk environmental DNA or RNA by sequencing. Meta-omic methods are particularly useful when applied to prokaryotes but they can also be used to monitor eukaryotes as well. It is believed that approximately only 5% of prokaryotic taxa can be characterized by culture-dependent methods. But meta-omics applications have demonstrated that only 1% of microbiota is culturable out of the vast varieties found on earth. Environmental microbiota helps in successful restoration of ecosystems; as they mediate many ecosystem restoration activities, including nutrient cycling, obligate and facultative symbioses (Livne-Luzon et al., 2017), carbon sequestration (Clemmensen et al., 2013), and human health (Mills et al., 2019; Robinson et al., 2018). The success of ecological restoration is monitored by the quantification and characterization of this microbiota by using meta-omics tools. For example, the recovery of soil fungi and bacteria following a restoration intervention in a post-agricultural context can be assessed using meta-barcoding, which measures alpha diversity and beta diversity (Yan et al., 2018) and derives restoration indicator taxa (Liddicoat et al., 2019) from DNA sequence abundances of marker genes. Metagenomics can be used to monitor the abundance of functional genes of soil microbiota. This approach to monitoring could be particularly useful in situations that demand a detailed understanding of nutrient changes with restoration, such as in post-agricultural settings where restoration activities often aim to reduce the soil fertility legacy of agricultural fertilizer use. Meta-transcriptomic analysis of RNA has been used to characterize the function of genes and the role of rare microorganisms in nitrogen fixation and sulfur oxidation in acid mine drainage (Hua et al., 2015). This approach holds promise for assessing the recovery of microbiota in the restoration of contaminated land, like the post-mining sites, and in turn the recovery potential of the entire ecosystem.

3.8. Plant Tissue Culture and Preservation Techniques

In recent decades, the loss of genetic diversity has led to a global biodiversity crisis. Tissue culture emerged as a promising technique for the restoration of genetic resources. Tissue culture has been practiced widely for germplasm conservation of medicinal herbs, plantation crops as well as endangered plants across the globe. Gene banks store the germplasm as well as ensure the global supply of endangered and commercially valuable species. Micropropagation is a promising technique to produce progeny plants of genetically modified plants, seedless plants, and plants that do not respond to vegetative reproduction in high count in less time.

In the recent past many developments have been done in this field to ensure successful preservations. Cryopreservation is a technique for viable freezing of genetic material at ultra-low temperatures in liquid nitrogen. In vitrification process, the highly concentrated solutions of cryoprotectants are rapidly cooled, under very low temperatures, which makes them viscous and gradually solidify without forming ice (Rall & Fahy, 1985). It is a very simple process and does not involve any programmable freezing equipment. However, ultra-cooling of some cryoprotectants may cause cell toxicity. The simple freezing method was first used for cell culture protection of *Citrus sinensis* by Sakai et al., (1991). In this process, a mixture of 3 M glycerol and 0.4 M sucrose is used as a cryoprotectant. Here the sample is loaded in a transparent straw and simply frozen at −30°C followed by immersion in liquid nitrogen. This process not only increases the possible viability of cell cultures but is also technologically accessible and affordable. Through synthetic seed technology also known as encapsulation technology, somatic embryos are coated in alginate beads for storage and exchange of germplasm. Further, Fabre & Dereuddre (1990) combined the encapsulation and dehydration process to ensure more successful storage and high survival. Synthetic seeds maintain genetical fidelity with increased competence of *in vitro* propagation (Nyende et al., 2003). In the cryo-selection technique, the immature embryo derived from the callus of nonhardy plants can be frozen in liquid N_2 in the absence of cryoprotectants. Cryo-selected cell lines exhibit increased tolerance towards freezing (up to −12°C) as well as allow the selection of genetic variants. Hardening at the callus level also brings up more tolerant plant variants. This technique does not involve freezing, cooling rates, and thawing regimes.

3.9. Microbial Bioremediation

Bioremediation has proved effective in cleaning contaminated soil and water making it suitable for reuse. Microbial metabolism can change organic contaminants into forms that can be processed by the earth's biogeochemical cycles. The genetic make-up of several bacterial and fungal strains that metabolize chemical degradation of hazardous environmental contaminants is well investigated which involves different metabolic processes such as the breakdown, transformation, and accumulation of toxic substances as depicted in Figure 2. Bioremediation carried out by marine microorganisms has been promising in lowering the level of hazardous compounds. An essential step in this process is the capacity of microbes to directly consume various harmful compounds found in the soil and then biotransform those chemicals into products that are neither toxic to the environment nor to other living beings (Smith et al., 2015). The biodegradation of total petroleum hydrocarbons in a slurry phase bioreactor was recently studied using old garbage. To accelerate degradation, several environmental parameters must be altered to encourage microbial growth and metabolism (Chen et al., 2019). Marine microorganisms are well-suited for bioremediation because they can quickly adapt to a variety of toxic conditions. Terrestrial organisms are less equipped than marine microbes to adapt to the varying conditions of the ocean environment. Bioremediation mostly works by way of oxidation-reduction reactions (redox), in which one chemical species contributes an electron to another that absorbs it. Bioremediations are mostly carried out under aerobic and anaerobic conditions. In aerobic bioremediation molecular oxygen is the final electron acceptor for organic compounds such as petroleum, phenols, polyaromatic hydrocarbons, and other contaminants. Under ideal conditions, the biodegradation rates of low to moderate molecular weight aliphatic, alicyclic, and aromatic compounds can be very high. It is possible to physically aerate the soil above the water table by employing a number of methods, including land farming, composting, and bioventing (Garcia Frutos et al., 2010; Norris, 1993). Anaerobic bioremediation is the process of reducing levels of background electron acceptors such as oxygen (O), nitrate (NO_3), ferric ion (Fe^{3+}), etc., and promoting the biochemical reduction of pollutants that have been oxidized. These include chlorinated cyclic hydrocarbons (CT, CF), chlorinated ethane (PCE, TCE, DCE, and VC), and nitrate, perchlorate, RDX, TNT, and chlorinated compounds (TC, DCA) (Gangar et al., 2019).

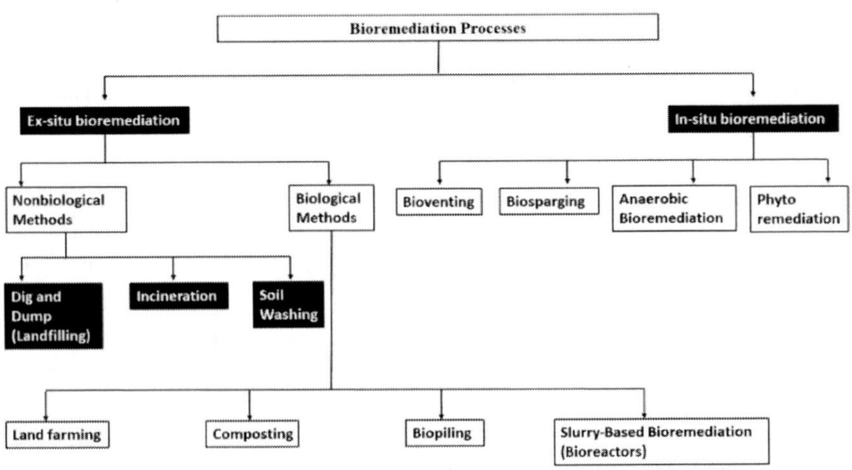

Figure 2. Processes involved in bioremediation.

3.10. Biodegradation of Xenobiotic Compounds

Biodegradation (Bioremediation) is an important biological and eco-friendly remedy for cleaning up the xenobiotic. Microbial biodegradation compiles three stages, namely, mineralization, biotransformation, and co-metabolism which work untidily as depicted in Figure 3. However, there are several xenobiotic compounds that have the capacity to avoid biodegradation or proceed with partial biodegradation (Singh, 2017).

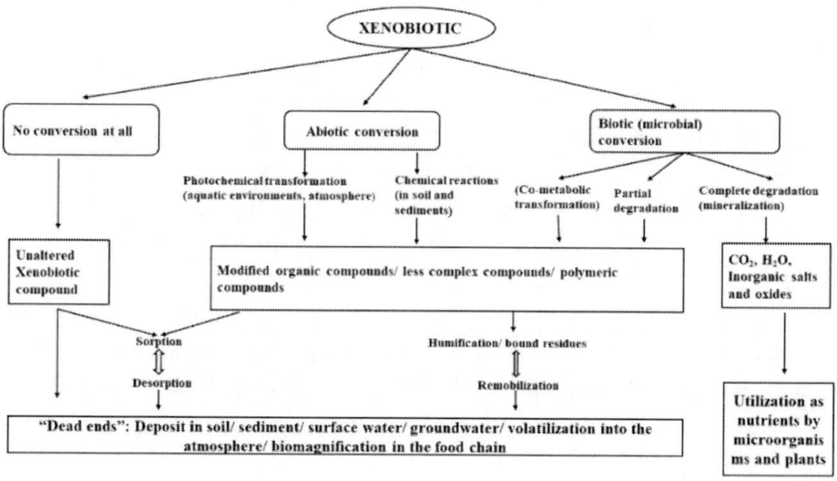

Figure 3. Possible environmental fate of a xenobiotic compound.

3.11. Application of CRISPR Cas9 Technology in Bioremediation

Recent advancements in biotechnology have made new opportunities for research in genetic engineering with respect to biodegradation. The idea of engineering a microbial metabolism to break down toxic compounds is an appealing possibility for remediating heavily polluted areas. A unique method known as gene editing allows for DNA manipulation using molecular scissors, a designed nuclease. The use of molecular scissors is extremely widespread in a variety of studies on plants, animals, and microbes. The bioremediation processes, such as the removal of xenobiotics and the transformation of harmful molecules into less toxic ones, can be significantly improved by the use of gene editing techniques (Basu et al., 2018; Hussain et al., 2018). Rachel Carson's "Silent Spring" (1962) was the impetus for research into toxic mineralization, which began in the 1950s. For the first time, CRISPR/Cas and its related proteins are being investigated as a method of genome editing. Genome alteration in microorganisms was previously accomplished by homologous recombination, group II intron retro homing, and multiplex automated genome engineering. All of these tools, meanwhile, were laborious and time-consuming.

CRISPR/Cas, which stands for clustered regularly interspaced short palindromic repeats in the defense mechanism found in bacteria and archaea against invading viruses. It has recently been engineered into a powerful gene editing tool in many organisms like *Escherichia coli* (Jiang et al., 2013), *Saccharomyces cerevisiae* and *Streptomyces* spp. (Cobb et al., 2015; DiCarlo et al., 2013), *Tatumella citrea*, Enterobacteriaceae (Jiang et al., 2015), and in many microbes of agricultural importance like *Bacillis subtilis* and *Bacillus mycoides*, fungal pathogens i.e., *Neurospora crassa*, *Mycelioptthora heterotalica*, *Aspergillus niger*, and *Aspergillus oryzae*, etc. (Muñoz et al., 2019). CRISPR- aided phytoremediation could be used to directly impart the desired set of instructions to the plant. It is a next-generation, programmable technique for high-throughput genetic manipulation. The accessibility of codon-optimized Cas9 versions has created new opportunities for CRISPR-Cas9 genome editing in a variety of plants (Kaur et al., 2019).

CRISPR-Cas is the most effective and productive gene-editing tool (McMahon et al., 2018). It is cheap, simple, and easy compared to other genome editing tools (Ju et al., 2018). It evaluates the gene interaction and thoroughly studies the genetic and phenotypic relationship along with the gene knock-out system (Vander Sluis et al., 2018). Various computational methods are an integrative approach for improving the bioremediation process. Contemporary enzyme-based mechanisms can be achieved by *in-silico* approaches as it is feasible to understand complex metabolic pathways of biodegradation and bioremediation (Liu et al., 2018). Various tools like flux balance analysis (FBA), metabolic flux analysis (MFA), and metabolic pathway analysis (MPA) are mostly used tools for stoichiometric analysis of metabolic networks (Gonzalez-Garcia et al., 2017). Flux can be designated as the flow of material with the edges carrying a value (Gomez & Barton, 2018). Knowledge of flux can be used to alter biological process dynamics through metabolic engineering (Cuperlovic-Culf, 2018). The gene of degrading bacteria can be manipulated, resulting in enhanced consumption of pesticide compounds (Sulpice & McKeown, 2015). To predict and understand the molecular level interactions of pesticidal compounds with bioremediating and degrading microbes, structure–activity relationship (QSAR) and 3DQSAR are used (Dreher et al., 2018). 3DQSAR also helps in studying the atomic-level interactions with ligands and compounds. These are used to study the biomagnification of pesticides (DDT) in marine and terrestrial ecosystem. Another tool, OptKnock is to get gene knock outs as well as to incorporate the genes encoding the novel enzymes for bioremediation.

3.12. Phytoremediation

There is no single bioremediation technique that can serve as a "silver bullet" to restore polluted environments. Bioremediations can either be carried out ex-situ (Landfilling, Incineration, Soil washing) or *in situ* (bioventing, phytoremediation). Through the use of genetic engineering and biotechnological techniques, plants can be genetically manipulated for enhanced phytoremediation which can extract, sequester, and detoxify a variety of xenobiotic as well as other contaminants and significantly clean up polluted soil, water, or air. The fundamental idea behind phytoremediation is that plants are cultivated in contaminated areas, and as they mature, they remove different types of contaminants and concentrate them in biomass (bioextraction) (Meagher, 2000).

Phytoremediation is the process of removing elemental pollutants, such as toxic heavy metals and radionuclides (McGrath & Zhao, 2003), and organic pollutants such as polychlorinated biphenyls (PCBs); polycyclic aromatic hydrocarbons (PAHs) such as benzopyrene; nitroaromatics such as trinitrotoluene (TNT); and linear halogenated hydrocarbons such as trichloroethylene (TCE) (Dietz & Schnoor, 2001) from the environment. There are some plants in the kingdom Plantae that are known to accumulate a large concentration of metals. Some of the hyper-accumulating or metal-resistant species are *Silene vulgaris*, *Arabidopsis halleri*, *Alyssum lesbiacum*, and *Brassica* spp. (Pająk et al., 2017, 2018) (Figure 4). Phytoremediation is focused on providing an alternative path of biological remediation and maintaining an ecological balance in an anthropogenically disturbed site. Pollutants can be remediated in plants through several natural processes, including adsorption, transport, and translocation. The degradation of endogenous toxic organics or their sequestration in vacuoles also protects plants from toxic xenobiotics (Dhankher et al., 2012; Meagher, 2000).

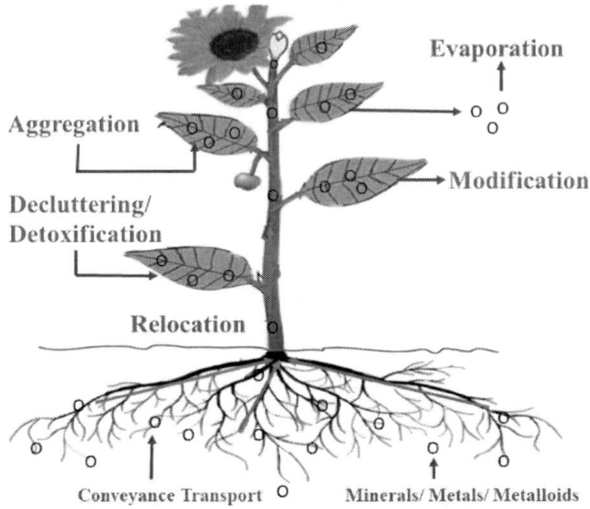

Figure 4. Illustration of the steps of phytoremediation, including the movement, accumulation, volatilization, and detoxification of pollutants.

Heavy metal accumulation is a persisting problem in the environment as it is toxic and nonbiodegradable, it enters the food chain and through biomagnification, it gets accumulated in the animal body. This is becoming a threat to human health as well as the ecosystem. Heavy metals like Cu, Fe, Mn, Ni, and Zn are much necessary for physio-chemical processes in plants but get toxic when accumulated in excess (Cempel & Nikel, 2006). Remediation of bioaccumulates is necessary so that the toxic heavy metals cannot enter the terrestrial, atmospheric, and aquatic environments. Metal ion transporters are responsible for translocating heavy metals. Metal transporters have been classified into several families like ZIP, HMAs, MTPs, and NRAMPs. HMA4 overexpression promoted metal tolerance by increasing Cd and Zn efflux from the root symplasm into the xylem vessels. Organic acids and amino acids also act as metal ligands by chelating heavy metal ions. Citrate mediates the chelation of Fe and Ni in the xylem (Lee et al., 1977), Ni is chelated by histidine (Krämer et al., 1996). Detoxification of heavy metals is the prerequisite for the initiation of phytoremediation (Thakur et al., 2016). There are two defense strategies that are implemented by plants i.e., avoidance and tolerance, which helps them to maintain the cellular concentrations of heavy metals below the toxicity threshold levels (Hall, 2002). Tolerance is mainly achieved through chelation by the complexation of heavy metal ions with ligands. Chelation results in the reduction of concentrations of free metal ions. Many organic and inorganic ligands in the cytoplasm help in heavy metal chelation. In the rhizospheric region, the root exudates act as heavy metal ligands to form stable heavy metal complexes (Dalvi & Bhalerao, 2013). The metal ions get chelated by altering the pH of the rhizosphere lessening the bioavailability and toxicity of the heavy metals (Dalvi & Bhalerao, 2013).

Conclusion and Future Prospects

Biodiversity is sensitive to even small changes in the earth's climate and needs sustainable methods for its conservation to maintain a continuous supply of ecosystem resources in its quality, value, and diversity. The protection of biodiversity has become a crucial and indispensable need. It is therefore important to understand and study the impacts of climate change on biodiversity and its mitigation strategies. Climate change altering ecosystem productivity and services may have many unforeseen consequences due to their strong interconnected network. Efficient high-output technologies used in plant germplasm collections, *in vitro* regeneration, molecular breeding, genome editing and big data management skills, have proved to be helpful in conservation and next-generation breeding unwrapping a new horizon.

References

Anderson, J. T., Inouye, D. W., McKinney, A. M., Colautti, R. I., & Mitchell-Olds, T. (2012). Phenotypic plasticity and adaptive evolution contribute to advancing flowering phenology in response to climate change. *Proceedings. Biological Sciences*, *279*(1743), 3843–3852. https://royalsocietypublishing.org/doi/10.1098/rspb.2012.1051.

Arroyo, M. T. K., Marquet, P., Marticorena, C., Simonetti, J., Cavieres, L., Squeo, F., & Rossi, R. (2004). Chilean winter rainfall – Valdivian Rainforest. In R. A. Mittermeier, P Robles, M. Hoffmann, J. Pilgrim, T. Brooks, C. Mittermeier, J. Lamoreux y G.A.B. Da Fonseca. (Eds.), *Hotspot revisited: Earth's biologically richest and most endangered terrestrial ecoregions* (pp. 99–103). CEMEX.

Awasthi, A., Uniyal, S. K., Rawat, G. S., & Rajvanshi, A. (2003). Forest resource availability and its use by the migratory villages of Uttarkashi, Garhwal Himalaya (India). *Forest Ecology and Management*, *174*(1–3), 13–24. https://doi.org/10.1016/S0378-1127(02)00026-9.

Bader, M. K. F., Ehrenberger, W., Bitter, R., Stevens, J., Miller, B. P., Chopard, J., Rüger, S., Hardy, G. E., Poot, P., Dixon, K. W., Zimmermann, U., & Veneklaas, E. J.

(2014). Spatio-temporal water dynamics in mature Banksia menziesii trees during drought. *Physiologia Plantarum*, *152*(2), 301–315. https://doi.org/10.1111/ppl.12170.

Bandelj, D., Jakše, J., & Javornik, B. (2004). Assessment of genetic variability of olive varieties by microsatellite and AFLP markers. *Euphytica*, *136*(1), 93–102. https://doi.org/10.1023/B:EUPH.0000019552.42066.10.

Barbraud, C., Weimerskirch, H., Bost, C. A., Forcada, J., Trathan, P., & Ainley, D. (2008). Are king penguin populations threatened by Southern Ocean warming? *Proceedings of the National Academy of Sciences of the United States of America*, *105*(26), E38–E39. https://doi.org/10.1073/pnas.0802013105.

Barrows, G., Sexton, S., & Zilberman, D. (2014). Agricultural biotechnology: The promise and prospects of genetically modified crops. *Journal of Economic Perspectives*, *28*(1), 99–120. https://doi.org/10.1257/jep.28.1.99.

Basu, S., Rabara, R. C., Negi, S., & Shukla, P. (2018). Engineering PGPMOs through gene editing and systems biology: A solution for phytoremediation? *Trends in Biotechnology*, *36*(5), 499–510. https://doi.org/10.1016/j.tibtech.2018.01.011.

Beniston, M. (2003). Climatic Change in Mountain Regions: A Review of Possible Impacts. In: Diaz, H.F. (eds) Climate Variability and Change in High Elevation Regions: Past, Present & Future. *Advances in Global Change Research*, vol 15. Springer, Dordrecht. https://doi.org/10.1007/978-94-015-1252-7_2.

Bhutiyani, M. R., Kale, V. S., & Pawar, N. J. (2010). Climate change and the precipitation variations in the northwestern Himalaya: 1866–2006. *International Journal of Climatology*, *30*(4), 535–548. https://doi.org/10.1002/joc.1920.

Boarman, W. I., & Sazaki, M. (1996). Highway mortality in desert tortoises and small vertebrates: Success of barrier fences and culverts. In G. J. Evink, P. Garrett, D. Zeigler & J. Berry (Eds.), *Trends in addressing transportation related wildlife mortality. Proceedings of the Transportation Related Wildlife Mortality Seminar*: Tallahassee, FL (pp. 169–173). Environmental Management Office, Department of Transportation.

Banjoko, B. (2014). Environmental Pharmacology – An Overview. *Pharmacology and Therapeutics*. http://dx.doi.org/10.5772/57473.

Breed, M. F., Harrison, P. A., Blyth, C., Byrne, M., Gaget, V., Gellie, N. J. C., Groom, S. V. C., Hodgson, R., Mills, J. G., Prowse, T. A. A., Steane, D. A., & Mohr, J. J. (2019). The potential of genomics for restoring ecosystems and biodiversity. *Nature Reviews. Genetics*, *20*(10), 615–628. https://doi.org/10.1038/s41576-019-0152-0.

Brimner, T. A., Gallivan, G. J., & Stephenson, G. R. (2005). Influence of herbicide-resistant canola on the environmental impact of weed management. *Pest Management Science*, *61*(1), 47–52. https://doi.org/10.1002/ps.967.

Bulucea, C. A., Rosen, M. A., Mastorakis, N. E., Bulucea, C. A., & Brindusa, C. C. (2012). Approaching resonant absorption of environmental xenobiotics harmonic oscillation by linear structure *Sustainability*, *4*(4), 561–573. https://doi.org/10.3390/su4040561.

Bureau of Land Management. (1990). *The California desert conservation area plan*. United States Department of the Interior. CDCA_Plan.pdf (blm.gov).

Camac, J. S., Williams, R. J., Wahren, C. H., Jarrad, F., Hoffmann, A. A., & Vesk, P. A. (2015). Modeling rates of life form cover change in burned and unburned alpine heathland subject to experimental warming. *Oecologia*, *178*(2), 615–628. https://doi.org/10.1007/s00442-015-3261-2.

Carson, R. (1962) *Silent Spring*. Houghton Miffl in, Boston, Massachusetts.

Cempel, M., & Nikel, G. J. P. J. S. (2006). Nickel: A review of its sources and environmental toxicology. *Polish Journal of Environmental Studies*, *15*(3), 375-382.

Challis, A., Stevens, J. C., Mcgrath, G., & Miller, B. P. (2016). Plant and environmental factors associated with drought-induced mortality in two facultative phreatophytic trees. *Plant and Soil*, *404*(1–2), 157–172. https://doi.org/10.1007/s11104-016-2793-5.

Chen, F., Li, X., Zhu, Q., Ma, J., Hou, H., & Zhang, S. (2019). Bioremediation of petroleum-contaminated soil enhanced by aged refuse. *Chemosphere*, *222*, 98–105. https://doi.org/10.1016/j.chemosphere.2019.01.122.

Clemmensen, K. E., Bahr, A., Ovaskainen, O., Dahlberg, A., Ekblad, A., Wallander, H., Stenlid, J., Finlay, R. D., Wardle, D. A., & Lindahl, B. D. (2013). Roots and associated fungi drive long-term carbon sequestration in boreal forest. *Science*, *339*(6127), 1615–1618. https://www.science.org/doi/10.1126/science.1231923.

Cobb, R. E., Wang, Y., & Zhao, H. (2015). High-efficiency multiplex genome editing of *Streptomyces* species using an engineered CRISPR/Cas system. *ACS Synthetic Biology*, *4*(6), 723–728. https://doi.org/10.1021/sb500351f.

Collen, B., & Nicholson, E. (2014). Conservation. Taking the measure of change. *Science (New York, N.Y.)*, *346*(6206), 166–167. https://www.science.org/doi/10.1126/science.1255772.

Cuklev, F., Fick, J., Cvijovic, M., Kristiansson, E., Förlin, L., & Larsson, D. G. (2012). Does ketoprofen or diclofenac pose the lowest risk to fish? *Journal of Hazardous Materials*, *229–230*, 100–106. https://doi.org/10.1016/j.jhazmat.2012.05.077.

Cuperlovic-Culf, M. (2018). Machine learning methods for analysis of metabolic data and metabolic pathway modeling. *Metabolites*, *8*(1), 4. https://doi.org/10.3390/metabo8010004.

Dalvi, A. A., & Bhalerao, S. A. (2013). Response of plants towards heavy metal toxicity: An overview of avoidance, tolerance and uptake mechanism. *Annals of Plant Sciences*, *2*(9), 362–368.

de Souza Machado, A. A., Kloas, W., Zarfl, C., Hempel, S., & Rillig, M. C. (2018). Microplastics as an emerging threat to terrestrial ecosystems. *Global Change Biology*, *24*(4), 1405–1416. https://doi.org/10.1111/gcb.14020.

Dhankher, O. P., Li, Y., Rosen, B. P., Shi, J., Salt, D., Senecoff, J. F., Sashti, N. A., & Meagher, R. B. (2002). Engineering tolerance and hyperaccumulation of arsenic in plants by combining arsenate reductase and γ-glutamylcysteine synthetase expression. *Nature Biotechnology*, *20*(11), 1140–1145. https://doi.org/10.1038/nbt747.

Dhankher, O. P., Pilon-Smits, E. A., Meagher, R. B., & Doty, S. (2012). Biotechnological approaches for phytoremediation. In *Plant biotechnology and agriculture* (pp. 309–328). Academic Press. https://doi.org/10.1016/B978-0-12-381466-1.00020-1.

DiCarlo, J. E., Norville, J. E., Mali, P., Rios, X., Aach, J., & Church, G. M. (2013). Genome engineering in *Saccharomyces cerevisiae* using CRISPR-Cas systems. *Nucleic Acids Research*, *41*(7), 4336–4343. https://doi.org/10.1093/nar/gkt135.

Diekmann, O. E., Gouveia, L., Perez, J. A., Gil-Rodriguez, C., & Serrão, E. A. (2010). The possible origin of Zostera noltii in the Canary Islands and guidelines for restoration. *Marine Biology*, *157*(9), 2109–2115. https://doi.org/10.1007/s00227-010-1467-8.

Dietz, A. C., & Schnoor, J. L. (2001). Advances in phytoremediation. *Environmental Health Perspectives*, *109*(Suppl 1) (Suppl. 1), 163–168. https://doi.org/10.1289/ehp.01109s1163.

Dimri, A. P., & Dash, S. K. (2012). Wintertime climatic trends in the western Himalayas. *Climatic Change*, *111*(3–4), 775–800. https://doi.org/10.1007/s10584-011-0201-y.

Doty, S. L., Shang, T. Q., Wilson, A. M., Tangen, J., Westergreen, A. D., Newman, L. A., Strand, S. E., & Gordon, M. P. (2000). Enhanced metabolism of halogenated hydrocarbons in transgenic plants containing mammalian cytochrome P450 2E1. *Proceedings of the National Academy of Sciences of the United States of America*, *97*(12), 6287–6291. https://doi.org/10.1073/pnas.97.12.6287.

Dreher, J., Scheiber, J., Stiefl, N., & Baumann, K. (2018). XMAP-an interpretable alignment-free four-dimensional quantitative structure-activity relationship technique based on molecular surface properties and conformer ensembles. *Journal of Chemical Information and Modeling*, *58*(1), 165–181. https://doi.org/10.1021/acs.jcim.7b00419.

Ducklow, H. W., Doney, S. C., & Steinberg, D. K. (2009). Contributions of long-term research and time-series observations to marine ecology and biogeochemistry. *Annual Review of Marine Science*, *1*(1), 279–302. https://doi.org/10.1146/annurev.marine.010908.163801.

Dyurgerov, M. B., & Meier, M. F. (2005). *Glaciers and the changing Earth system: A 2004 snapshot*, *58*. Institute of Arctic and Alpine Research, University of Colorado. ISSN 0069-6145.

Eriksson M., Fang J., Dekens J. (2008). How does climate change affect human health in the Hindu Kush-Himalaya region. *Regional Health Forum* 12, 11–15.

Faast, R., & Facelli, J. M. (2007). Investigation of processes leading to the decline of South Australia's Caladenia species. *Lankesteriana*, *7*(1–2), (1–2). https://doi.org/10.15517/lank.v7i1-2.19517.

Fabre, J., & Dereuddre, J. (1990). *Encapsulation-dehydration: A new approach to cryopreservation of Solanum shoot-tips*. Cryo-Letters, 11, pp. 413-426.

French, C. E., Rosser, S. J., Davies, G. J., Nicklin, S., & Bruce, N. C. (1999). Biodegradation of explosives by transgenic plants expressing pentaerythritol tetranitrate reductase. *Nature Biotechnology*, *17*(5), 491–494. https://doi.org/10.1038/8673.

Gangar, T., Bhardwaj, K. K. and Gupta, R. (2019). Microbes and processes in bioremediation of soil. *Microbes and Enzymes in Soil Health and Bioremediation*, pp. 11-37. https://doi.org/10.1007/978-981-13-9117-0_2.

Gao, L., Tang, S., Zhuge, L., Nie, M., Zhu, Z., Li, B., & Yang, J. (2012). Spatial genetic structure in natural populations of Phragmites australis in a mosaic of saline habitats in the Yellow River Delta, China. *PLOS ONE*, *7*(8), e43334. https://doi.org/10.1371/journal.pone.0043334.

Garcia Frutos, F. J. G., Escolano, O., García, S., Babín, M., & Fernández, M. D. (2010). Bioventing remediation and ecotoxicity evaluation of phenanthrene-contaminated soil. *Journal of Hazardous Materials*, *183*(1–3), 806–813. https://doi.org/10.1016/j.jhazmat.2010.07.098.

Genin, D., & Badan-Dangon, A. (1991). Goat herbivory and plant phenology in a Mediterranean shrubland of northern Baja California. *Journal of Arid Environments*, *21*(1), 113–121. https://doi.org/10.1016/S0140-1963(18)30733-X.

Godefroid, S., Piazza, C., Rossi, G., Buord, S., Stevens, A. D., Aguraiuja, R., Cowell, C., Weekley, C. W., Vogg, G., Iriondo, J. M., Johnson, I., Dixon, B., Gordon, D., Magnanon, S., Valentin, B., Bjureke, K., Koopman, R., Vicens, M., Virevaire, M., & Vanderborght, T. (2011). How successful are plant species reintroductions? *BiologicalConservation*, *144*(2), 672–682. https://doi.org/10.1016/j.biocon.2010.10.003.

Gomez, J. A., & Barton, P. I. (2018). Dynamic flux balance analysis using DFBAlab. In *Metabolic network reconstruction and modeling* (pp. 353–370). Humana Press. https://doi.org/10.1007/978-1-4939-7528-0_16.

Gonzalez-Garcia, R. A., Aispuro-Castro, R., Salgado-Manjarrez, E., Aranda-Barradas, J., & Garcia-Peña, E. I. (2017). Metabolic pathway and flux analysis of H2 production by an anaerobic mixed culture. *International Journal of Hydrogen Energy*, *42*(7), 4069–4082. https://doi.org/10.1016/j.ijhydene.2017.01.043.

Gradinger, R. (2009). Sea-ice algae: Major contributors to primary production and algal biomass in the Chukchi and Beaufort Seas during May/June 2002. *Deep Sea Research Part II*, *56*(17), 1201–1212. https://doi.org/10.1016/j.dsr2.2008.10.016.

Guerrero, P. C., León-Lobos, P. E. D. R. O. and Squeo, F. A. (2008). Prioritization of Endemic Species Present in the Atacama Region: Taxonomic Uniqueness and Degrees of Endemism. *Red Book of Native Flora and Priority Sites for Conservation: Atacama Region*, pp. 339-346.

Guhathakurta, P., & Rajeevan, M. (2008). Trends in the rainfall pattern over India. *International Journal of Climatology*, *28*(11), 1453–1469. https://doi.org/10.1002/joc.1640.

Gupta, R. K. (1978). Impact of human influences on the vegetation of the Western Himalaya. *Vegetatio*, *37*(2), 111–118. https://doi.org/10.1007/BF00126834.

Hall, J. L. (2002). Cellular mechanisms for heavy metal detoxification and tolerance. *Journal of Experimental Botany*, *53*(366), 1–11. https://doi.org/10.1093/jexbot/53.366.1.

Hirschi, K. D., Korenkov, V. D., Wilganowski, N. L., & Wagner, G. J. (2000). Expression of Arabidopsis CAX2 in tobacco. Altered metal accumulation and increased manganese tolerance. *Plant Physiology*, *124*(1), 125–133. https://doi.org/10.1104/pp.124.1.125.

Hochachka, P. W., & Somero, G. N. (2002). *Biochemical adaptation: Mechanism and process in physiological evolution*. Oxford University Press.

Hoegh-Guldberg, O., & Bruno, J. F. (2010). The impact of climate change on the world's marine ecosystems. *Science, 328*(5985), 1523–1528. https://www.science.org/doi/10.1126/science.1189930.

Hoegh-Guldberg, O., Mumby, P. J., Hooten, A. J., Steneck, R. S., Greenfield, P., Gomez, E., Harvell, C. D., Sale, P. F., Edwards, A. J., Caldeira, K., Knowlton, N., Eakin, C. M., Iglesias-Prieto, R., Muthiga, N., Bradbury, R. H., Dubi, A., & Hatziolos, M. E. (2007). Coral reefs under rapid climate change and ocean acidification. *Science, 318*(5857), 1737–1742. https://www.science.org/doi/10.1126/science.1152509.

Hoffmann, M., Hilton-Taylor, C., Angulo, A., Böhm, M., Brooks, T. M., Butchart, S. H., Carpenter, K. E., Chanson, J., Collen, B., Cox, N. A., Darwall, W. R., Dulvy, N. K., Harrison, L. R., Katariya, V., Pollock, C. M., Quader, S., Richman, N. I., Rodrigues, A. S., Tognelli, M. F., Stuart, S. N. (2010). The impact of conservation on the status of the world's vertebrates. *Science, 330*(6010), 1503–1509. https://www.science.org/doi/10.1126/science.1194442.

Hua, Z. S., Han, Y. J., Chen, L. X., Liu, J., Hu, M., Li, S. J., Kuang, J. L., Chain, P. S., Huang, L. N., & Shu, W. S. (2015). Ecological roles of dominant and rare prokaryotes in acid mine drainage revealed by metagenomics and metatranscriptomics. *ISME Journal, 9*(6), 1280–1294. https://doi.org/10.1038/ismej.2014.212.

Huerta Lwanga, E. H., Gertsen, H., Gooren, H., Peters, P., Salánki, T., van der Ploeg, M., Besseling, E., Koelmans, A. A., & Geissen, V. (2017). Incorporation of microplastics from litter into burrows of Lumbricus terrestris. *Environmental Pollution, 220*(A), 523–531. https://doi.org/10.1016/j.envpol.2016.09.096.

Hunter, R. (1991). *Bromus* invasions on the Nevada Test Site: Present status of *B. rubens* and *B. tectorum* with notes on their relationship to disturbance and altitude. *Great Basin Naturalist, 51*, 176–182.

Hussain, B., Lucas, S. J., & Budak, H. (2018). CRISPR/Cas9 in plants: At play in the genome and at work for crop improvement. *Briefings in Functional Genomics, 17*(5), 319–328. https://doi.org/10.1093/bfgp/ely016.

IPCC (Intergovernmental Panel on Climate Change) (2007): The physical science basis. *Climate Change.* In: *Contribution of Working Group I to the Fourth assessment report of the Intergovernmental Panel on Climate Change* (pp. 235–336). Cambridge University Press. http://www.ipcc.ch/S.

IPCC (Intergovernmental Panel on Climate Change) (2001): Impacts, adaptation, and vulnerability. Contribution of working Group II to the third assessment report of the Intergovernmental Panel on Climate Change: Cambridge, *Climate Change.* Cambridge University Press. http://www.ipcc.ch/S.

IPCC (Intergovernmental Panel on Climate Change) (2013): The Physical Science Basis. Contribution of Working Group I to the Fifth Assessment Report of the Intergovernmental Panel on Climate Change. Summary for policy makers. In *Climatic Change*. Cambridge University Press, (1–32).

Jiang, W., Bikard, D., Cox, D., Zhang, F., & Marraffini, L. A. (2013). RNA-guided editing of bacterial genomes using CRISPR-Cas systems. *Nature Biotechnology, 31*(3), 233–239. https://doi.org/10.1038/nbt.2508.

Jiang, Y., Chen, B., Duan, C., Sun, B., Yang, J., & Yang, S. (2015). Multigene editing in the *Escherichia coli* genome via the CRISPR-Cas9 system. *Applied and Environmental Microbiology*, *81*(7), 2506–2514. https://doi.org/10.1128/AEM.04023-14.

Jin, Q., Waters, D., Cordeiro, G. M., Henry, R. J., & Reinke, R. F. (2003). A single nucleotide polymorphism (SNP) marker linked to the fragrance gene in rice (Oryza sativa L.). *Plant Science*, *165*(2), 359–364. https://doi.org/10.1016/S0168-9452(03)00195-X.

Ju, X. D., Xu, J., & Sun, Z. S. (2018). CRISPR editing in biological and biomedical investigation. *Journal of Cellular Biochemistry*, *119*(1), 52–61. https://doi.org/10.1002/jcb.26154.

Kaur, R., Sinha, K., & Bhunia, R. K. (2019). Can wheat survive in heat? Assembling tools towards successful development of heat stress tolerance in *Triticum aestivum L. Molecular Biology Reports*, *46*(2), 2577–2593. https://doi.org/10.1007/s11033-019-04686-x.

Khraiwesh, B., Zhu, J. K., & Zhu, J. (2012). Role of miRNAs and siRNAs in biotic and abiotic stress responses of plants. *Biochimica et Biophysica Acta*, *1819*(2), 137–148. https://doi.org/10.1016/j.bbagrm.2011.05.001.

Krämer, U., Cotter-Howells, J. D., Charnock, J. M., Baker, A. J. M., & Smith, J. A. C. (1996). Free histidine as a metal chelator in plants that accumulate nickel. *Nature*, *379*(6566), 635–638. https://doi.org/10.1038/379635a0.

Kumar, A., & Ram, J. (2005). Anthropogenic disturbances and plant biodiversity in forests of Uttaranchal, central Himalaya. *Biodiversity and Conservation*, *14*(2), 309–331. https://doi.org/10.1007/s10531-004-5047-4.

Le Page, Y., Vosges, M., Servili, A., Brion, F., & Kah, O. (2011). Neuroendocrine effects of endocrine disruptors in teleost fish. *Journal of Toxicology and Environmental Health. Part B, Critical Reviews*, *14*(5–7), 370–386. https://doi.org/10.1080/10937404.2011.578558.

Lechner, A., & Ramler, D. (2015). The discharge of certain amounts of industrial microplastic from a production plant into the River Danube is permitted by the Austrian legislation. *Environmental Pollution*, *200*, 159–160. https://doi.org/10.1016/j.envpol.2015.02.019.

Lee, J., Reeves, R. D., Brooks, R. R., & Jaffré, T. (1977). Isolation and identification of a citrato-complex of nickel from nickel-accumulating plants. *Phytochemistry*, *16*(10), 1503–1505. https://doi.org/10.1016/0031-9422(77)84010-7.

Liddicoat, C., Weinstein, P., Bissett, A., Gellie, N. J. C., Mills, J. G., Waycott, M., & Breed, M. F. (2019). Can bacterial indicators of a grassy woodland restoration inform ecosystem assessment and microbiota-mediated human health? *Environment International*, *129*, 105–117. https://doi.org/10.1016/j.envint.2019.05.011.

Lightfoot, D. C. and Whitford, W. G. (1991). Productivity of creosotebush foliage and associated canopy arthropods along a desert roadside. *American Midland Naturalist*, pp. 310-322. https://doi.org/10.2307/2426235.

Liu, S., Song, H., Wei, S., Yang, F., & Li, X. (2014). Bio-cathode materials evaluation and configuration optimization for power output of vertical subsurface flow constructed wetland—Microbial fuel cell systems. *Bioresource Technology*, *166*, 575–583. https://doi.org/10.1016/j.biortech.2014.05.104.

Liu, W. X., Liu, W. H., Wu, J., Gao, A. N., & Li, L. H. (2010). Analysis of genetic diversity in natural populations of Psathyrostachys huashanica Keng using microsatellite (SSR) markers. *Agricultural Sciences in China*, *9*(4), 463–471. https://doi.org/10.1016/S1671-2927(09)60118-8.

Liu, Z., Liu, Y., Zeng, G., Shao, B., Chen, M., Li, Z., Jiang, Y., Liu, Y., Zhang, Y., & Zhong, H. (2018). Application of molecular docking for the degradation of organic pollutants in the environmental remediation: A review. *Chemosphere*, *203*, 139–150. https://doi.org/10.1016/j.chemosphere.2018.03.179.

Livne-Luzon, S., Ovadia, O., Weber, G., Avidan, Y., Migael, H., Glassman, S. I., Bruns, T. D., & Shemesh, H. (2017). Small-scale spatial variability in the distribution of ectomycorrhizal fungi affects plant performance and fungal diversity. *Ecology Letters*, *20*(9), 1192–1202. https://doi.org/10.1111/ele.12816.

Lovich, J. E., & De Gouvenain, R. C. (1998). Saltcedar invasion in desert wetlands of the southwestern United States: ecological and political implications. *Proceedings California Exotic Pest Council*, *4*, 45-55.

Mao, D., Tian, Y., Wang, Z., Jia, M., Du, J., & Song, C. (2021). Wetland changes in the Amur River Basin: Differing trends and proximate causes on the Chinese and Russian sides. *Journal of Environmental Management*, *280*, 111670. https://doi.org/10.1016/j.jenvman.2020.111670.

Mason, S. A., Garneau, D., Sutton, R., Chu, Y., Ehmann, K., Barnes, J., Fink, P., Papazissimos, D., & Rogers, D. L. (2016). Microplastic pollution is widely detected in US municipal wastewater treatment plant effluent. *Environmental Pollution*, *218*, 1045–1054. https://doi.org/10.1016/j.envpol.2016.08.056.

Matusick, G., Ruthrof, K. X., Brouwers, N. C., Dell, B., & Hardy, G. S. J. (2013). Sudden forest canopy collapse corresponding with extreme drought and heat in a Mediterranean-type eucalypt forest in southwestern Australia. *European Journal of Forest Research*, *132*(3), 497–510. https://doi.org/10.1007/s10342-013-0690-5.

McGrath, S. P., & Zhao, F. J. (2003). Phytoextraction of metals and metalloids from contaminated soils. *Current Opinion in Biotechnology*, *14*(3), 277–282. https://doi.org/10.1016/s0958-1669(03)00060-0.

McMahon, M. A., Prakash, T. P., Cleveland, D. W., Bennett, C. F., & Rahdar, M. (2018). Chemically modified Cpf1-CRISPR RNAs mediate efficient genome editing in mammalian cells. *Molecular Therapy*, *26*(5), 1228–1240. https://doi.org/10.1016/j.ymthe.2018.02.031.

Meagher, R. B. (2000). Phytoremediation of toxic elemental and organic pollutants. *Current Opinion in Plant Biology*, *3*(2), 153–162. https://doi.org/10.1016/s1369-5266(99)00054-0.

Mills, J. G., Brookes, J. D., Gellie, N. J. C., Liddicoat, C., Lowe, A. J., Sydnor, H. R., Thomas, T., Weinstein, P., Weyrich, L. S., & Breed, M. F. (2019). Relating urban biodiversity to human health with the 'holobiont' concept. *Frontiers in Microbiology*, *10*, 550. https://doi.org/10.3389/fmicb.2019.00550.

Min, X. Y., Zhang, Z. S., Liu, Y., Wei, X., Liu, Z. P., Wang, Y. R., & Liu, W. X. (2017). Genome-wide development of microRNA-based SSR markers in *Medicago truncatula* with their transferability analysis and utilization in related legume species. *International Journal of Molecular Sciences*, *18*(11), 2440. https://doi.org/10.3390/ijms18112440.

Mondal, T. K., Ganie, S. A., & Debnath, A. B. (2015). Identification of novel and conserved miRNAs from extreme halophyte, Oryza coarctata, a wild relative ofrice. *PLOSONE*, *10*(10),e0140675. https://doi.org/10.1371/journal.pone.0140675.

Mukhopadhyay, B. (2013). Global warming–A threat to the planet. *American International Journal of Biology*, *1*, 29–34.

Muñoz, I. V., Sarrocco, S., Malfatti, L., Baroncelli, R., & Vannacci, G. (2019). CRISPR-Cas for fungal genome editing: A new tool for the management of plant diseases. *Frontiers in Plant Science*, *10*, 135. https://doi.org/10.3389/fpls.2019.00135.

Nautiyal, M. C., Nautiyal, B. P., & Prakash, V. (2004). Effect of grazing and climatic changes on alpine vegetation of Tungnath, Garhwal Himalaya, India. *Environmentalist*, *24*(2), 125–134. https://doi.org/10.1007/s10669-004-4803-z.

Neilson, J. R., & Sharp, P. A. (2008). Small RNA regulators of gene expression. *Cell*, *134*(6),899–902. https://doi.org/10.1016/j.cell.2008.09.006.

Nelson, D. E., Repetti, P. P., Adams, T. R., Creelman, R. A., Wu, J., Warner, D. C., Anstrom, D. C., Bensen, R. J., Castiglioni, P. P., Donnarummo, M. G., Hinchey, B. S., Kumimoto, R. W., Maszle, D. R., Canales, R. D., Krolikowski, K. A., Dotson, S. B., Gutterson, N., Ratcliffe, O. J., & Heard, J. E. (2007). Plant nuclear factor Y (NF-Y) B subunits confer drought tolerance and lead to improved corn yields on water-limited acres. *Proceedings of the National Academy of Sciences of the United States of America*, *104*(42), 16450–16455. https://doi.org/10.1073/pnas.0707193104.

NOAA, U. S. National Oceanic and Atmospheric Administration (2009). *State of the Climate Report*.

Nogués-Bravo, D., Araújo, M. B., Errea, M. P., & Martínez-Rica, J. P. (2007). Exposure of global mountain systems to climate warming during the 21st Century. *Global Environmental Change*, *17*(3–4), 420–428. https://doi.org/10.1016/j.gloenvcha.2006.11.007.

Norris, R. D. (1993). *Handbook of bioremediation*. CRC Press. https://doi.org/10.1201/9780203712764.

Nyende, A. B., Schittenhelm, S., Mix-Wagner, G., & Greef, J. M. (2003). Production, storability, and regeneration of shoot tips of potato (Solanum tuberosum L.) encapsulated in calcium alginate hollow beads. *In Vitro Cellular and Developmental Biology – Plant*, *39*(5), 540–544. https://doi.org/10.1079/IVP2003442.

O'Connor, M. I., Piehler, M. F., Leech, D. M., Anton, A., & Bruno, J. F. (2009). Warming and resource availability shift food web structure and metabolism. *PLOS Biology*, *7*(8), e1000178. https://doi.org/10.1371/journal.pbio.1000178.

Paetkau, D., VÃ¡zquez-DomÃ-nguez, E., Tucker, N. I. J., & Moritz, C. (2009) Monitoring movement into and through a newly planted rainforest corridor using genetic analysis of natal origin. *Ecological Management and Restoration*, *10*(3),210–216. https://doi.org/10.1111/j.14428903.2009.00490.x.

Pająk, M., Błońska, E., Szostak, M., Gąsiorek, M., Pietrzykowski, M., Urban, O., & Derbis, P. (2018). Restoration of vegetation in relation to soil properties of spoil heap heavily contaminated with heavy metals. *Water, Air, & Soil Pollution*, *229*, 1-15. https://doi.org/10.1007/s11270-018-4040-6.

Pająk, M., Halecki, W., & Gąsiorek, M. (2017). Accumulative response of Scots pine (*Pinus sylvestris* L.) and silver birch (*Betula pendula* Roth) to heavy metals enhanced by Pb-Zn ore mining and processing plants: Explicitly spatial considerations of ordinary kriging based on a GIS approach. *Chemosphere*, *168*,851–859. https://doi.org/10.1016/j.chemosphere.2016.10.125.

Polovina, J. J., Howell, E. A., & Abecassis, M. (2008). Ocean's least productive waters are expanding. *Geophysical Research Letters*, *35*(3). https://doi.org/10.1029/2007GL031745.

Prose, D. V., Metzger, S. K., & Wilshire, H. G. (1987). Effects of substrate disturbance on secondary plant succession; Mojave Desert, California. *Journal of Applied Ecology*, *24*(1), 305–313. https://doi.org/10.2307/2403806.

Rall, W. F., & Fahy, G. M. (1985). Ice-free cryopreservation of mouse embryos at -196 degrees by vitrification. *Nature*, *313*(6003), 573–575. https://doi.org/10.1038/313573a0.

Ramey, R. R., Luikart, G., & Singer, F. J. (2000). Genetic bottlenecks resulting from restoration efforts: The case of bighorn sheep in Badlands National Park. *Restoration Ecology*, *8*(4S), 85–90. https://doi.org/10.1046/j.1526-100x.2000.80069.x.

Rejmánek, M., & Randall, J. M. (1994). Invasive alien plants in California: 1993 summary and comparison with other areas in North America. *Madrono*, 161-177.

Robbirt, K. M., Davy, A. J., Hutchings, M. J., & Roberts, D. L. (2011). Validation of biological collections as a source of phenological data for use in climate change studies: A case study with the orchid Ophrys sphegodes. *Journal of Ecology*, *99*(1), 235–241. https://doi.org/10.1111/j.1365-2745.2010.01727.x.

Robertson, B. C., & Gemmell, N. J. (2004). Defining eradication units to control invasive pests. *Journal of Applied Ecology*, *41*(6), 1042-1048. https://doi.org/10.1111/j.0021-8901.2004.00984x.

Robinson, J. M., Mills, J. G., & Breed, M. F. (2018). Walking ecosystems in microbiome-inspired green infrastructure: An ecological perspective on enhancing personal and planetary health. *Challenges*, *9*(2), 40. https://doi.org/10.3390/challe9020040.

Rossetto, M., Bragg, J., Kilian, A., McPherson, H., van der Merwe, M., & Wilson, P. D. (2019). Restore and renew: A genomics-era framework for species provenance delimitation. *Restoration Ecology*, *27*(3), 538–548. https://doi.org/10.1111/rec.12898.

Rugh, C. L., Wilde, H. D., Stack, N. M., Thompson, D. M., Summers, A. O., & Meagher, R. B. (1996). Mercuric ion reduction and resistance in transgenic Arabidopsis thaliana plants expressing a modified bacterial merA gene. *Proceedings of the National Academy of Sciences of the United States of America*, *93*(8), 3182–3187. https://doi.org/10.1073/pnas.93.8.3182.

Sakai, A., Kobayashi, S., & Oiyama, I. (1991). Cryopreservation of nucellar cells of navel orange (*Citrus sinensis* Osb.) by a simple freezing method. *Plant Science*, *74*(2), 243–248. https://doi.org/10.1016/0168-9452(91)90052-A.

Sakamoto, A., & Murata, N. (2002). The role of glycine betaine in the protection of plants from stress: Clues from transgenic plants. *Plant, Cell and Environment*, *25*(2),163-171. https://doi.org/10.1046/j.00168025.2001.00790.x.

Salazar Suazo, E. R., León Lobos, P. M., Muñoz Schick, C. and Rosas, C. (2006). *Ex situ conservation status of cultivated and wild plant genetic resources in Chile.*

Semwal, R. L., Nautiyal, S., Sen, K. K., Rana, U., Maikhuri, R. K., Rao, K. S., & Saxena, K. G. (2004). Patterns and ecological implications of agricultural land-use changes: A case study from central Himalaya, India. *Agriculture, Ecosystems and Environment*, *102*(1), 81–92. https://doi.org/10.1016/S0167-8809(03)00228-7.

Serra, A. A., Nuttens, A., Larvor, V., Renault, D., Couée, I., Sulmon, C., & Gouesbet, G. (2013). Low environmentally relevant levels of bioactive xenobiotics and associated degradation products cause cryptic perturbations of metabolism and molecular stress responses in *Arabidopsis thaliana*. *Journal of Experimental Botany*, *64*(10), 2753–2766. https://doi.org/10.1093/jxb/ert119.

Shanmugam, G., Sampath, S., Selvaraj, K. K., Larsson, D. J., & Ramaswamy, B. R. (2014). Non-steroidal anti-inflammatory drugs in Indian rivers. *Environmental Science and Pollution Research*, *21*, 921-931. https://doi.org/10.1007/s11356-013-1957-6.

Shrestha, A. B., Wake, C. P., Mayewski, P. A., & Dibb, J. E. (1999). Maximum temperature trends in the Himalaya and its vicinity: An analysis based on temperature records from Nepal for the period 1971–94. *Journal of Climate*, *12*(9),2775–2786. https://doi.org/10.1175/15200442(1999)012<2775:MTTITH>2.0.CO;2.

Sikandar, A., Shehzadi, K., Arshad, Q., & Munir, K. (2013). Phytoremediation: An analytical technique for the assessment of biodegradation of organic xenobiotic pollutants: A review. *International Journal of Sciences*, *4*(2), 2250–2253.

Sinclair, T. R., Purcell, L. C., & Sneller, C. H. (2004). Crop transformation and the challenge to increase yield potential. *Trends in Plant Science*, *9*(2), 70–75. https://doi.org/10.1016/j.tplants.2003.12.008.

Singh, I., Smita, S., Mishra, D. C., Kumar, S., Singh, B. K., & Rai, A. (2017). Abiotic stress responsive miRNA-target network and related markers (SNP, SSR) in Brassica juncea. *Frontiers in Plant Science*, *8*, 1943. https://doi.org/10.3389/fpls.2017.01943.

Singh, R. (2017). Biodegradation of xenobiotics-a way for environmental detoxification. *International Journal of Development Research*, *7*(1), 14082–14087.

Smith, E., Thavamani, P., Ramadass, K., Naidu, R., Srivastava, P., & Megharaj, M. (2015). Remediation trials for hydrocarbon-contaminated soils in arid environments: Evaluation of bioslurry and biopiling techniques. *International Biodeterioration and Biodegradation*, *101*, 56–65. https://doi.org/10.1016/j.ibiod.2015.03.029.

Sohoni, P., & Sumpter, J. P. (1998). Several environmental oestrogens are also anti-androgens. *Journal of Endocrinology*, *158*(3), 327–339. https://doi.org/10.1677/joe.0.1580327.

Song, W. Y., Sohn, E. J., Martinoia, E., Lee, Y. J., Yang, Y. Y., Jasinski, M., Forestier, C., Hwang, I., & Lee, Y. (2003). Engineering tolerance and accumulation of lead and cadmium in transgenic plants. *Nature Biotechnology*, *21*(8), 914–919. https://doi.org/10.1038/nbt850.

Sontakke, N. A., Singh, H. N., & Singh, N. (2010). Monitoring physiographic rainfall variation for sustainable Management of Water Bodies in India. *Natural and anthropogenic disasters: vulnerability, preparedness and mitigation*, 293-331. https://doi.org/10.1007/978-90-481-2498-5_14.

Spear, S. F., Crisafulli, C. M., & Storfer, A. (2012). Genetic structure among coastal tailed frog populations at Mount St. Helens is moderated by post-disturbance management. *Ecological Applications*, *22*(3), 856–869. https://doi.org/10.1890/11-0627.1.

Stone, R. (2008). Plant science. China plans $3.5 billion GM crops initiative. *Science*, *321*(5894),1279-1279. https://doi.org/10.1126/science.321.5894.1279.

Suding, K., Higgs, E., Palmer, M., Callicott, J. B., Anderson, C. B., Baker, M., Gutrich, J. J., Hondula, K. L., LaFevor, M. C., Larson, B. M. and Randall, A. (2015). Committing to ecological restoration. *Science*, *348*(6235), pp.638-640. https://www.science.org/doi/pdf/10.1126/science.aaa4216.

Sulpice, R., & McKeown, P. C. (2015). Moving toward a comprehensive map of central plant metabolism. *Annual Review of Plant Biology*, *66*, 187–210. https://doi.org/10.1146/annurev-arplant-043014-114720.

Supple, M. A., Bragg, J. G., Broadhurst, L. M., Nicotra, A. B., Byrne, M., Andrew, R. L., Widdup, A., Aitken, N. C., & Borevitz, J. O. (2018). Landscape genomic prediction for restoration of a Eucalyptus foundation species under climate change. *eLife*, *7*, e31835. https://doi.org/10.7554/eLife.31835.

Thakur, S., Singh, L., Wahid, Z. A., Siddiqui, M. F., Atnaw, S. M., & Din, M. F. M. (2016). Plant-driven removal of heavy metals from soil: Uptake, translocation, tolerance mechanism, challenges, and future perspectives. *Environmental Monitoring and Assessment*, *188*(4), 206. https://doi.org/10.1007/s10661-016-5211-9.

Tierra Madre Consultants. (1991). *Biological assessment for Lancaster City and planning area: Relative density surveys for desert tortoises and cumulative human impact evaluations for Mohave ground squirrel habitat*, Riverside, California.

Tong, L., Xu, X., Fu, Y., & Li, S. (2014). Wetland changes and their responses to climate change in the "three-river headwaters" region of China since the 1990s. *Energies*, *7*(4), 2515–2534. https://doi.org/10.3390/en7042515.

Vandepitte, K., Gristina, A. S., De Hert, K., Meekers, T., Roldán-Ruiz, I., & Honnay, O. (2012). Recolonization after habitat restoration leads to decreased genetic variation in populations of a terrestrial orchid. *Molecular Ecology*, *21*(17),4206-4215. https://doi.org/10.1111/j.1365294X.2012.05698.x.

VanderSluis, B., Costanzo, M., Billmann, M., Ward, H. N., Myers, C. L., Andrews, B. J., & Boone, C. (2018). Integrating genetic and protein–protein interaction networks maps a functional wiring diagram of a cell. *Current Opinion in Microbiology*, *45*, 170–179. https://doi.org/10.1016/j.mib.2018.06.004.

Varshney, R. K., Bansal, K. C., Aggarwal, P. K., Datta, S. K., & Craufurd, P. Q. (2011). Agricultural biotechnology for crop improvement in a variable climate: Hope or hype? *Trends in Plant Science*, *16*(7), 363–371. https://doi.org/10.1016/j.tplants.2011.03.004.

Vinnikov, K. Y., Groisman, P. Y., & Lugina, K. M. (1990). Empirical data on contemporary global climate changes (temperature and precipitation). *Journal of Climate*, *3*(6), 662–677. https://doi.org/10.1175/15200442(1990)003<0662:EDOCGC>2.0.CO;2.

Wahren, C. -H., Camac, J. S., Jarrad, F. C., Williams, R. J., Papst, W. A., & Hoffmann, A. A. (2013). Experimental warming and long-term vegetation dynamics in an alpine heathland. *Australian Journal of Botany*, *61*(1), 36–51. https://doi.org/10.1071/BT12234.

Wang, T., O'Neill, G. A., & Aitken, S. N. (2010). Integrating environmental and genetic effects to predict responses of tree populations to climate. *Ecological Applications*, *20*(1), 153–163. https://doi.org/10.1890/08-2257.1.

Wang, Z., Song, K., Ma, W., Ren, C., Zhang, B., Liu, D., Chen, J. M., & Song, C. (2011). Loss and fragmentation of marshes in the Sanjiang Plain, Northeast China, 1954–2005. *Wetlands*, *31*(5), 945–954. https://doi.org/10.1007/s13157-011-0209-0.

Wikramanayake, E., Dinerstein, E., Allnut, T., Loucks, C. and Wettengel, W., 1998. A biodiversity assessment and gap analysis of the Himalayas. *World Wildlife Fund-US, Conservation Science Program/UNDP Report.*

Williams, C. (1996). Combatting marine pollution from land-based activities: Australian initiatives. *Ocean & Coastal Management*, *33*(1–3), 87–112. https://doi.org/10.1016/S0964-5691(96)00046-4.

Wilshire, H., & Prose, D. (1987). Wind energy development in California, USA. *Environmental Management*, *11*(1), 13–20. https://doi.org/10.1007/BF01867174.

Wu, F., Zhang, D., Ma, J., Luo, K., Di, H., Liu, Z., Zhang, J., & Wang, Y. (2016). Analysis of genetic diversity and population structure in accessions of the genus Melilotus. *Industrial Crops and Products*, *85*, 84–92. https://doi.org/10.1016/j.indcrop.2016.02.055.

Yan, D., Mills, J. G., Gellie, N. J. C., Bissett, A., Lowe, A. J., & Breed, M. F. (2018). High-throughput eDNA monitoring of fungi to track functional recovery in ecological restoration. *Biological Conservation*, *217*, 113–120. https://doi.org/10.1016/j.biocon.2017.10.035.

Yang, J., Pedlar, J. H., McKenney, D. W., & Weersink, A. (2015). The development of universal response functions to facilitate climate-smart regeneration of black spruce and white pine in Ontario, Canada. *Forest Ecology and Management*, *339*, 34–43. https://doi.org/10.1016/j.foreco.2014.12.001.

Young, J. A. (1991). Tumbleweed. *Scientific American*, *264*(3), 82–87. https://doi.org/10.1038/scientificamerican0391-82.

Zink, T. A., Allen, M. F., Heindl-Tenhunen, B., & Allen, E. B. (1995). The effect of a disturbance corridor on an ecological reserve. *Restoration Ecology*, *3*(4), 304–310. https://doi.org/10.1111/j.1526-100X.1995.tb00098.x.

Zhang J, Duan Z, Jahufer Z, An S, Wang Y (2014) Stress-inducible expression of a Cleistogenes songorica ALDH gene enhanced drought tolerance in transgenic Arabislopsis thaliana. *Plant Omics* 7:438.

Zhang, Z., Min, X., Wang, Z., Wang, Y., Liu, Z., & Liu, W. (2017). Genome-wide development and utilization of novel intron-length polymorphic (ILP) markers in

Medicago sativa. *Molecular Breeding*, *37*(7), 1–8. https://doi.org/10.1007/s11032-017-0659-z.

Zhu, D., Chen, Q. L., An, X. L., Yang, X. R., Christie, P., Ke, X., Wu, L. H., & Zhu, Y. G. (2018). Exposure of soil collembolans to microplastics perturbs their gut microbiota and alters their isotopic composition. *Soil Biology and Biochemistry*, *116*, 302–310. https://doi.org/10.1016/j.soilbio.2017.10.027.

Zink, T. A., Allen, M. F., Heindl-Tenhunen, B., & Allen, E. B. (1995). The effect of a disturbance corridor on an ecological reserve. *Restoration Ecology*, *3*(4), 304–310. https://doi.org/10.1111/j.1526-100X.1995.tb00098.x.

Zonunsanga, R., Rao, C. B., & Rinawma, P. (2014). Degradation of Land and Forest Resources: The Story of Shifting Cultivation and Loss of Biodiversity in North-East India. https://doi.org/10.1007/978-4-431-54838-6_20.

Chapter 7

Phytoremediation: A Tool for the Reclamation of Pesticides and Pharmaceutical Contaminated Ecosystems

Bhawana Raj[1]
Rachna[1]
and Umesh Kumar Singh[1,2,*]
[1]Department of Environmental Science, Central University of South Bihar, Gaya, Bihar, India
[2]Centre of Environmental Studies, University of Allahabad, Prayagraj, Uttar Pradesh, India

Abstract

Pharmaceutical compounds and pesticides are commonly found both in the aquatic and agricultural environments because of human activities and the associated discharge of wastewater effluents and runoff to the environment. Despite the low concentrations of pharmaceutical compounds and pesticides usually found, the continuous introduction into the environment from different pathways makes them 'pseudo-persistent.' The toxicity and persistence of the pesticide and pharmaceutical residues released by various anthropogenic activities are hazardous because of their toxicity and persistent nature. Due to their potential for toxicity, mutagenicity, carcinogenicity, and genotoxicity, they are chemicals of concern. The existing conventional clean-up techniques are typically costly and may only change one harmful contaminant into another. Ecological-based strategies are preferable to

* Corresponding Author's Email: umeshsingh@cusb.ac.in.

In: Conserving, Halting and Reinstating Degraded Natural Resources …
Editors: Rajani Srivastava and Pankaj Kumar
ISBN: 979-8-89113-067-8

conventional methods in the treatment of chemical effluents. Phytoremediation is proven to be green technology that employs plants to clean up the waters, soils, and sediments. To remove contaminants from wastewater effluents and aquatic ecosystems, aquatic macrophytes have been widely utilized. For optimal benefits, genetic engineering can be integrated with phytoremediation by having a thorough understanding of its mechanism. The key parts of the environmental ecosystem examined in this chapter are soil, water, and sediments. The chapter discusses the use of various aquatic macrophytes and plants for the remediation of pharmaceutical compounds and pesticides present in soil/sediments and water, and it also points out the research gaps that must be filled to find practical and sound ecological solutions for the remediation of polluted environments for sustainable development. The current study aims to map the current understanding of pharmaceutical chemicals and pesticide interactions with plants regarding absorption and the application of plant-based systems for phytoremediation.

Keywords: phytoremediation, pesticides, pharmaceutical compounds, ecosystem

1. Introduction

The hazardous synthetic chemicals that were produced to assist mankind have now become a concern for living organisms. A very significant quantity of organic and inorganic compounds enters the environment and poses risk to its structure and functioning and the most often identified and detected compounds are pharmaceuticals and pesticides. In recent decades, pharmaceuticals and pesticides have grabbed a lot of attention among researchers due to their ubiquitous presence in the water and soil (Petrovic et al., 2010; Van De Steene et al., 2010). However, these pesticides and pharmaceuticals are found largely in the ecosystem but they differ in their origin and the pathways leading them to the ecosystem (Proia et al., 2013). Pesticides are thought to be very helpful for crops and in-house cleaning products because of their pest management qualities and are utilized by farmers regularly in the field and are mostly found their way into the environment through practices. Just like other synthetic chemicals pesticides and pharmaceuticals have its limitation and disadvantages. Although pharmaceutical concentration in the environment has been generally observed lower than pesticides (Petrovic et al., 2005) yet, it is equally harmful to human and animal health. Pesticide use has increased dramatically because of the industrialization of agriculture. These two substances have regularly been found in the ecosystem at concentrations ranging from nanograms and micrograms. Despite normally being in the allowed concentration, their concentrations frequently go beyond the allowed limit. Even if these pesticides are present in the atmosphere in less than the permitted amount, their synergistic effects may still have serious negative impacts on health. These compounds may or may not be degraded or metabolized after usage and are persistent for a longer period. The conventional method of extraction and analysis of these pollutants is mostly expensive, less effective, and does not have an environmentally friendly approach. As more environmental awareness and innovation increase, there is a greater need for environmentally friendly, cost-effective, and more efficient options for the remediation of these pollutants. Phytoremediation is one such method that has attracted a lot of attention in recent years due to its environmentally friendly approach and advantages over traditional practices.

1.1. Phytoremediation

The capacity of plants to remove contaminants from the ecosystem is known as phytoremediation (Figure 1).

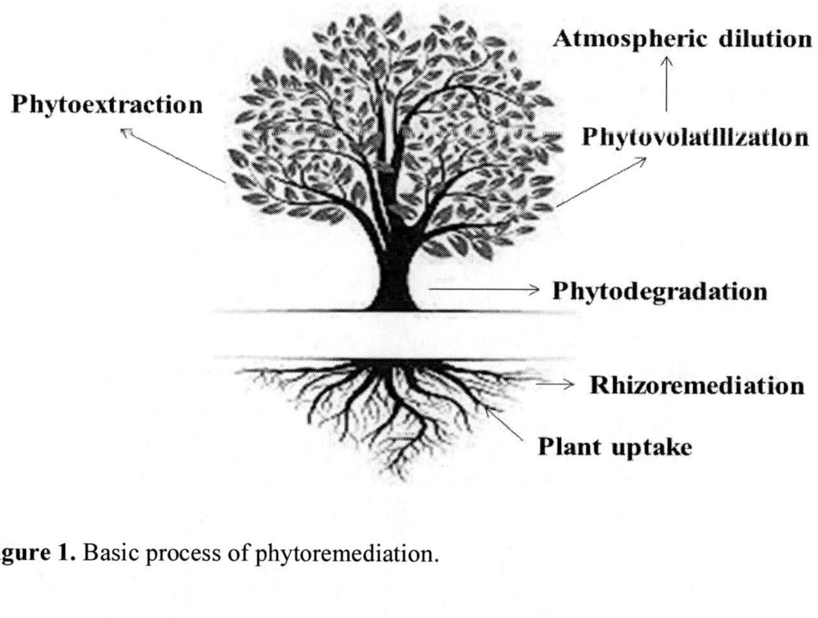

Figure 1. Basic process of phytoremediation.

Some phytoremediation processes, including phytostabilization, phytotransformation, phytovolatilization, phytofiltration, and phytoextraction, result in the production of non-toxic compounds.

2. Pesticides

Over the past few decades, the usage of pesticides has multiplied greatly. Many of them are "persistent," in nature thus they last for a longer period without dissipating into safer component elements and eventually find their way into our food chain along with their presence in the water and the air. In the study conducted by Chang (2018), residues of organochlorine pesticides majorly DDT were detected in the fish. The highest detection was of dieldrin followed by cis-chlordane and hexachlorobenzene. Another study has found chlorpyrifos, DDT (dichlorodiphenyltrichloroethane), and Ƴ- HCH (hexachlorocyclohexane) in higher amounts in the milk samples collected from Punjab, India. β-endosulfan, endosulfan sulfate, cypermethrin, cyhalothrin, fenvalerate, deltamethrin, Malathion, profenofos, and ethion were also detected (Bedi et al., 2015). Out of 37.7%, cabbage, legumes, and leaf mustard samples have shown 17.2%, 18.9%, and 17.2% pesticide residues exceeding the maximum residue limits (MRLs) (Chen et al., 2011). In the 1940s, the first synthetic pesticides became commercially accessible, and in the 1960s, concerns about their potential health impacts emerged. A pesticide is a harmful chemical substance or mixture of toxic compounds that are applied to houses, agricultural fields, and other areas to control insects, weeds, rodents, fungus, and other dangerous pests in the form of sprays, poisons, and powders. The major types of pesticides are organochlorines, carbamates, organophosphates, pyrethroids, and neonicotinoids (Mahmood et al., 2013). Depending on the crop stage, formulation, the intended target, and weather circumstances, pesticides in the agricultural field may be sprayed as liquid sprays, sometimes incorporated or injected into the soil, or applied as granules or as a seed treatment. The toxicity of pesticides is not only limited to the target pests and often exerts toxicity to the non-target species causing damage to the biodiversity and the ecological health of the animals and human beings. The "Dirty Dozens" are a group of 12 compounds that have been designated as the most hazardous and persistent substances by the Stockholm Convention on Persistent Organic Pollutants (POPs). Ten of these 12 compounds fall under the pesticide category (Gilden et al., 2010). Pesticides possess numerous health and environmental problems due to accidental or deliberate exposure and cause higher health risks to both humans and animals (Cooper & Dobson, 2007). Organochlorines had a significant negative influence on top predators in terrestrial food chains, such as birds of prey, and accumulated in the adipose tissues of both people and animals. They also function as endocrine disruptors and are passed to neonates through breast milk fat (Carvalho, 2017) and exhibit carcinogenicity, genotoxicity, and immunotoxicity.

Pesticide exposures are causing several symptoms, including dizziness, headaches, nausea, vomiting, diarrhea, blindness, itchy skin, rash, and even death (Fang et al., 2020). Each pesticide has a different amount of toxicity and the potential to risk human health. For example, both acetaminophen and morphine are analgesics but morphine stronger analgesic (Carvalho, 2017). In the past few decades, agricultural output has increased significantly as a result of the population boom. The curve of population growth has only been rising during the past few decades. According to current growth rates, the world's population will be between 9.4 and 10 billion by 2050, up from 1.5 billion in 1900 and 6.1 billion in 2000 (UN, 2015). This increase in the population can only be supported by more food production that can be achieved with the help of pesticides. The utilization of pesticides is not going to halt anytime soon. Innovative solutions must be created to handle this issue, but there have only been a few studies in the area. Phytoremediation is one of the options for remediation, and due to its emphasis on environmental friendliness, it has received greater attention.

2.1. Phytoremediation of Pesticides

Over the past, many physiochemical technologies have been utilized to remove pesticides from the environment. However, conventional technologies are not environmentally friendly and sometimes have harmful effects. Phytoremediation is a green technology that has attracted much interest recently because of its environmentally friendly mechanisms. Phytoremediation depends on many factors such as mode of application, soil type, climatic factors, and the species of plants. Many plant species are efficient in removing pesticides. It has also been shown that aquatic plants have better remediating potentials and more biomass. For example, water hyacinths can remove organ phosphorus like ethion from water, and over time, the quantity of ethion inside the plant lowers, indicating the metabolic capacity of hyacinth plants. Other plants, such as E. crassipes, P. stratiotes, and algae, in comparison to their natural degradation in aquatic environments, needed a short duration of pesticide bioaccumulation. When examined with organochlorines and pyrethroid, it is discovered that P. stratiotes have a stronger ability for bioaccumulation than E. crassipes (Riaz et al., 2017). Some wild plants have also shown the ability to accumulate pesticides like DDT metabolites that can further help in the degradation or removal of these contaminants. Sunflower has demonstrated outstanding achievements in soil pesticide reduction because of their high phytoextraction percentage, supported by strong biomass output (Mitton et al., 2016). When combined with other technologies or techniques, these remediating plant species have demonstrated exceptional potential. According to a 2015 research by. Romeh (2015), adding synthetic surfactants like Tween 80 to Glycine max L. and Plantago major L. improved the removal of azoxystrobin from the polluted soil. Another such example is the integration of nanotechnology and phytoremediation which has degraded 81.20% of endosulfan in a short period of 28 days in natural environmental conditions (Jesitha and Harikumar, 2018). It is important to recognize and select the best plant species, which may be done by comprehending the physiological and biochemical traits. A robust bioaccumulation capacity and the ability to transport toxins to the shoots, where additional pollutants may be absorbed, are necessary for a perfect phytoremediator. Ideal phytoremediators plants make the process more effective; For instance, cucurbits have shown better uptake of dieldrin and endrin than jute and hence will be more effective in the phytoremediation of dieldrin and endrin (Otani et al., 2007). Along with certain early reactions including reduction, oxidation, and hydrolysis, the three primary stages in the cleanup of pesticides are transformation, conjugation, and sequestration. The most common transformation processes in the metabolism of pesticides are oxidation reactions. Depending on the pesticide and plant type, and due to saturation over time, a plant's capacity to translocate and metabolize a pesticide might change. Atrazine, azoxystrobin, and imidacloprid are all metabolized differently by I. versicolor, A. virginicus, and P. virginity. The plant's capacity to absorb and swiftly metabolize the pesticides may be the reason for the lower concentration retrieved. Switchgrass has demonstrated an excellent atrazine metabolic capability. Most of the 14C that the grass has absorbed is transformed into metabolites that go from the sand via the roots to the leaves. Activated carbon made from Eichhornia crassipes is considered a bio-adsorbent and has been used in the removal of malathion (Malik, 2007).

Pollutant availability and level of toxicity are key factors in phytoremediation effectiveness. Pollutant availability will have an immediate impact on soil-root transfer, and toxicity that might restrict plant development, which will ultimately have an impact on the absorption and translocation processes. Each plant species may, however, modify the soil-root environment, interact specifically with the rhizosphere, and adjust its tolerance level to the environment. Table 1 listed selected plant species having potential for phytoremediation. Organophosphorus pesticides do not show bioaccumulation variances in their log Kow range, contrary to the organochlorine pesticides, whose bioaccumulation statistics indicate that bioaccumulation is governed by their log Kow (Mercado -Borrayo et al., 2015). Pollutants must be safely sequestered or detoxified since they stress plants, which might affect their ability to maintain their membrane integrity or metabolic pathways. When under extreme and ongoing stress, reactive oxygen species (ROS) build up and cause damage like modifications to proteins and membranes. The oxidative stress in plant species may have increased protein oxidation, lipid peroxidation, and antioxidant system activation (Mitton et al., 2016). Malathion has a significant influence on plant DNA. Strong mutagenesis effects were visible in the DNA of plants with RAPD-PCR (random amplified polymorphic DNA-polymerase chain reaction) patterns. These mutagenic effects have a positive impact on the phytoremediation potential of plants. The phytoremediation of endosulfan does not affect the morphological structure of the plants and hence, the site contaminated with endosulfan can be suitable for phytoremediation (Mukherjee and Kumar, 2012).

Table 1. Phytoremediation potential of some of the plants

Plant	Pesticides	Reference
Ricinus communis L.	OCPs	Rissato et al., 2015
Juncus effusus	atrazine	Bouldin et al., 2006
Ludwigiapeploides	lambda-cyhalothrin	Bouldin et al., 2006
P. stratiotes	organochlorine and pyrethroid	Riaz et al., 2017
E. crassipes	organochlorine and pyrethroid	Riaz et al., 2017
Cichoriumintybus	DDT	Suresh et al., 2005
Brassica juncea	DDT	Suresh et al., 2005
A. calcarata	endosulfan	Jesitha & Harikumar, 2018
Cucurbita sp.	dieldrin and endrin	Otani et al., 2007
Glycine max L. and Plantago major L	azoxystrobin	Romeh, 2015
Withania somnifera	HCH	Abhilash and Singh, 2010

3. Pharmaceutical Compounds in an Ecosystem

In the last years, pharmaceutical compounds have been gaining prominence due to their potential negative effects on both the aquatic environment and human health (Halling-Sørensen et al., 1998). Pharmaceutical compounds are increasingly polluting the environment, especially the aquatic environment. These pollutants enter the environment because of pharmaceutical industry activities, and improper disposal of unused or expired drug wastes generated in hospitals and stock-raising farms (Wojcieszyńska et al., 2014). Most pharmaceutical compounds are not fully degraded after application and get into the environment in an unmovable or slightly modified form. Due to their stability, they are not eliminated by the sewage treatment plants, thus they can be unintentionally consumed by humans in tap water (Wojcieszyńska et al., 2014). Conventional wastewater treatment plants (WWTPs) are not specifically designed to eliminate pharmaceutical compounds. Thus, many of these pollutants can easily pass through the wastewater treatment processes due to their continuous introduction.

The removal technologies of pharmaceutical compounds use advanced technologies such as advanced oxidative processes, activated carbon adsorption, membrane filtration, and membrane bioreactors (Anjali & Shanthakumar, 2019; Morone et al., 2019). These technologies are efficient for the removal of these contaminates but are not used widely because of relatively expensive. As a result, there is a need for an alternative on-site treatment process for the removal of pharmaceutical compounds that have high efficiency of operation and maintenance at a reasonable cost. One of the most promising technologies is phytoremediation. Phytoremediation is a technology that utilizes plants and the associated rhizosphere microorganisms to remove, transform or contain toxic chemicals located in soils, sediments, groundwater, surface water, and even the atmosphere (Susarla et al., 2002). This is a cost-effective environmental restoration technique that is also eco-friendly.

The principal application of phytoremediation is for lightly contaminated soils and waters, where the material to be treated is at a shallow or medium depth and the area to be treated is large. Plants that can decontaminate soils do one or more of the following: (1) plant uptake of contaminant from soil particles or soil liquid into their roots; (2) bind the contaminant into their root tissue, physically or chemically; and (3) transport the contaminant from their roots into growing shoots and prevent or inhibit the contaminant from leaching out of the soil (Lee et al., 2003). Moreover, the plants should not only accumulate, degrade, or volatilize the contaminants but should also grow quickly in a range of different conditions and lend themselves to easy harvesting. If the plants are left to die *in situ*, the contaminants will return to the soil. So, for the complete removal of contaminants from an area, the plants must be cut and disposed of elsewhere in a non-polluting way.

3.1. Phytoremediation of Pharmaceutical Compounds in Soil and Water

Pharmaceutical compounds present in an ecosystem can cause a problem for soil biota, groundwater, and public health (Schaefer & Juliane, 2007). Pharmaceutical compounds can change soil and water chemistry and make them useless (Anderson & LaBelle, 2000; Obot et al., 2006; Okpokwasili & Odokuma,1990). Pharmaceutical residues, which are reversibly adsorbed to the soil, may be taken up by plants (Fatta-Kassinos et al., 2011) and, in this way, transferred to animals and humans. Water and soil pollution is one of the key challenges for a million people especially in a developing country (Landrigan et al., 2017). Nowadays this pollution growing unconscious release of wastewater in water bodies and disposal of waste on open land is a concern for the sustainable development of water and soil resources for society as well as the environment.

Some plants and aquatic macrophytes adapted for the remediation of pollutants in water and soil (Ekperusi et al., 2019). Some examples of plants used in phytoremediation practices are the following: water hyacinths (*Eichornia crassipes*); poplar trees (*Populus* spp.); forage kochia (*Kochia* spp.); alfalfa (*Medicago sativa*); Kentucky bluegrass (*Poa pratensis*); *Scirpus* spp., coontail (*Ceratophyllum demersum* L.); American pondweed (*Potamogeton nodosus*); and the emergent common arrowhead (*Sagittaria latifolia*) among others (Pereira et al., 2006). The types of macrophytes used for the remediation of pharmaceutical compounds in wastewater namely floating macrophytes, emergent macrophytes, and submerged macrophytes (Ekperusi et al., 2019).

(a) *Floating Macrophytes:* - The macrophytes that are exclusively found on the surface of water bodies and also in standing or slow-moving waters (Ekperusi et al., 2019). Lemna species, *Pistia stratiotes, Salvinia minimum, Landoltia punctate, Spirodela polyrhiza, Riccia fluitans, Wolffiella gladiata, Marsilea mutica, Hottonia inflata*, and *Azolla* species are a few examples. (USDA., 2018)

(b) *Emergent Macrophytes:* - The macrophytes that have roots attached to the substrate at the bottom of water bodies and the leaves grow to or above the surface of the water (Ekperusi et al., 2019). Examples include *Butomus umbellatus, Cyperus* spp.*, Diodia virginiana, Distichlis spicata, Hydrochloa caroliniensis, Imperatacylindrica, Iris virginica, Juncus* spp., *Justicia americana, Nasturtium officinale, Nuphar lutea, Nymphaea* spp., *Phragmites australis and Typha* spp. (USDA., 2018)

(c) *Submerged Macrophytes:* - The macrophytes that are permanently submerged in water and their root system is attached to the substrate but their leaves do not reach the surface of the water, unlike emergent macrophytes. They are enduringly submerged throughout the year or lifecycle. A few examples are *Ceratophyllum demersum, Egeria densa, Hydrilla verticillata, Hygrophila corymbosa, Myriophyllum aquaticum, Najas marina, Potamogeton natans, Ruppia maritima, Stuckenia pectinata, Vallisneria americana, and Elodea canadensis*. (Mahmood et al., 2013; USDA., 2018).

According to Anand et al., (2017), aquatic plants are essential in biological wastewater treatment systems because they help in the phytoremediation process through phytoextraction, rhizofiltration, phytovolatilization, phytodegradation, or Phytotransformation techniques. Many of the plants have remediation capacity of pharmaceutical compounds from water and sediments (Table 2). Floating aquatic plants have become more suitable for phytoremediation because of their availability, high yield, and ease of stocking and harvesting (Ekperusi et al., 2019). The submerged species appears to transform and/ or degrade the pharmaceutical compound while floating species accumulate high concentrations in their tissues (Kitamura et al., 2022). Floating aquatic plants grow in large quantities so their biomass is easy to recover and then can be used for bioenergy through direct combustion or for biogas or bioethanol production, avoiding the return of the contaminant to the environment.

The removal mechanisms of the different medicines were examined, including plant absorption, sorption, microbial degradation in plants, volatilization, and photodegradation (Amare et al., 2018). Pollutants eradication depends upon several factors such as duration of exposure, the concentration of pollutants, environmental factors (pH, temperature), and plant characteristics (species, root system, etc.) Physical characteristics, genetic adaptations, and metabolic and absorption capabilities of plants must be considered while choosing a plant for phytoremediation. A need for strong phytoremediation capacity is species with higher climate adaptability and higher biomass output (Misra and Misra, 2019).

Table 2. Removal efficiency of different plants for the remediation of pharmaceutical compounds from water and sediments

Pharmaceutical Compounds	Plants	Removal (%)	References
Acetaminophen (30–181 mg L^{-1})	Armoratia rusticana (horseradish)	100	Kotyza et al., (2010)
Acetaminophen (15–181 mg L^{-1})	Lupinus luteolus (pale yellow lupine)	100	Kotyza et al., (2010)
Ciprofloxacin (100 -1μgL^{-1})	Salvinia molesta	69 to 93	Kitamura et al., (2022)
Caffeine (10 μg L^{-1})	Elodea canadensis (aquatic plant)	99 ± 1	Matamoros et al., (2012)
Clofibric acid (20 μg L^{-1})	Typha spp.	80	Dordio et al., (2009)
Carbamazepine (0.5–2 mg L^{-1}	Typha spp.	82	Dordio et al., (2011)
Diclofenac (0.5–2 mg L^{-1})	Scirpus Validus	>98	Zhang et al., (2012)
Flumequine (0.050–1 mg L^{-1})	Lemna minor (aquatic plant)	96	Cascone et al., (2004)
Ibuprofen (0.03–30 mg L^{-1})	Populus nigra	100	Iori et al., (2012)
Naproxen (100 mg L^{-1}	Vetiver zizaniodes	81.5	Marsidi et al., (2016)
Oxytetracycline (0.5–2.0 mg L^{-1})	Pistia stratiotes (water lettuce	>90	Gujarathi et al., (2005)
Sulfapyridine (200 mg L^{-1}, wastewater	Zannichellia palustris (aquatic plant)	83 ± 5	Garcia-Rodríguez et al., (2013)
Sulfadimethoxine (12 mg L^{-1})	Hordeum vulgare (barley)	75	Ferro et al., (2010)
Sulfadimethoxine (50–450 mg L^{-1})	Azzola filicuolides (aquatic plant)	56.3–88.5	Forni et al., (2002)
Tetracycline (5–15 mg L^{-1})	Chrysopogon zizanioides	100	Datta et al., (2013)

Iatrou et al., (2017) have reported that *Lemna minor* which is the common duckweed removes four antimicrobials such as cefadroxil, metronidazole, trimethoprim, and sulfamethoxazole more efficiently. Authors mention that the duckweeds remove cefadroxil upto100%, metronidazole 96%, trimethoprim 59%, and sulfamethoxazole 73% within 24 days of periods. Reinhold et al., 2010 compared the two species of duckweeds such as *L. minor* and *L. punctate* for the phytoremediation of pharmaceutical compounds. According to authors *L. minor* and *L. punctata* showed similar remediation patterns for ibuprofen, fluoxetine, and triclosan within 9 days. According to research by Matamoros et al., 2012, the four aquatic plants *S. molesta, L. minor, C. demersum,* and *E. canadensis* effectively eliminate a variety of common medications, including diclofenac, triclosan, naproxen, ibuprofen, caffeine, and clofibric acid. They claimed that the four plants effectively removed diclofenac, triclosan, and caffeine with 81-99%, whereas clofibric acid was only eliminated with 6-23% efficiency. Of the four species, C. demersum was noted as being the most effective, followed by E. canadensis, L. minor, and S. molesta, which was the least effective. Many pharmaceuticals were successfully removed from wetlands by *Myriophyllum verticillate, Typha orientalis*, and *L. minor*, according to Zhou et al., (2018). Among the three macrophytes, *M. verticillatum* was the most effective and the least was *L. minor*. Kitamura et al., (2022) reported that *S. molesta* (floating) and *E. densa* (submerged) remove Ciprofloxacin from contaminated water. Both macrophytes species are efficient at reclaiming the antibiotic (>60%) even when at very high concentrations, such as those found in effluents from hospitals and pharmaceutical industries (concentrations varying from 2.3 to 341 micrograms per liter). Allam et al., (2015) investigated how well the duckweed (*Lemna gibba*) plant removed drugs including acetaminophen, diclofenac, and progesterone from drainage water at various beginning concentrations of 1, 50, 500, and 1000 micrograms per liter. *Vetiveria zizanioides*, a phytoremediator plant, proved successful in removing naproxen as a micropollutant from contaminated soil and water. Naproxen removal up to 81.5%, 64.5%, and 60.9% was achieved in 100, 300, and 500 mg/L naproxen concentrations, respectively. Phytostimulation and phytostabilization are the main processes which able to eliminate naproxen from wastewater and soil when bacteria are present in the soil around the roots of the plants (Marsidi et al., 2016). Hoang et al., (2013) reported the removal of pharmaceutical compounds from sediment by plants such as *Rhizophora apiculata* and *Acrostichum aureum* which remove ciprofloxacin and norfloxacin efficiently. Pharmaceutical substances can be absorbed by roots, transported to shoots, and metabolized by plants. Not all plant species have this potential, and the characteristics of the medicinal chemicals are crucial. Moreover, plants can also take up volatilized pharmaceutical compounds from the air through above-ground plant parts (Collins et al., 2011). According to the literature sunflower which is a c_3 plant and maize which is a C_4 plant can also be used for the phytoremediation of pharmaceutical compounds. According to Ryslava et al., (2015), maize can remove 10,11epoxy carbamazepine (a metabolite of carbamazepine) from polluted soils with up to 88% effectiveness in 96 hours. Sunflowers are also effective in removing carbamazepine. Datta et al., (2013) took a step toward a long-term goal of remediating tetracycline-contaminated soil and water using vetiver grass.

Future Perspectives

The phytoremediation approach can be a useful tool in the current situation to address the issue of pesticide and pharmaceutical residues. However, the traditional phytoremediation methods are not economical from the perspective of mass application because the naturally occurring hyperaccumulators are generally slow and produce relatively less biomass. The potential of native plants and their relationship with microbes in the clean-up of pharmaceuticals and pesticides has to be further investigated. It has been noted that the potential association of plants and microbes suitable for remediation has also encountered several difficulties, such as different pollutants other than the target pollutant affecting the contaminated sites, plants responding differently in water and soil, various environmental conditions of the experiments, and associated microorganisms affecting the removal, degradation, and metabolization of pollutants. A lot of information is lacking on the mobility and fate of pollutants in plant tissues. Future research is needed in this direction to fully understand the pathways and fate of pollutants in plant tissues. Understanding the plant's accumulative and degradative pathways in greater detail may help in a better understanding of the mechanisms in the plant and pave the way for genetic engineering. More pilot-scale research should be prioritized to supplement current information and better understand the limits of this technique. It is necessary to perform more thorough research on the interactions between soil, microbes, and plants to successfully translate bench- and pilot-scale findings to the field level. Understanding enzymatic processes have enhanced our understanding of pesticide science and plant and microbial biochemistry and physiology, and it now requires the integration of other technologies with phytoremediation to monitor the efficiency and overcome the limits. Knowledge of enzymatic processes, particularly ideas linked to pesticides' mechanism of action, resistance, selectivity, tolerance, and environmental fate needs to be studied in more detail. To make these technologies more efficient, time-saving, and financially viable, additional study is required in the field of genetic engineering to enhance the phytoremediation capabilities of transgenic plants and to comprehend the processes and efficiency of phytoremediation techniques. Research on the mutual symbiotic relationships between different plant or microbial activities may also be an intriguing direction to take in terms of boosting plant growth and, consequently, the effectiveness of phytoremediation. Another crucial area of study is advancement in the science of plant engineering, which gives plants their unique characteristics.

Conclusion

Management of water pollution, soil pollution, and air pollution from municipal, industrial, and agricultural activities are of critical concern to the government, policymakers, and the general public. Several gaps in phytoremediation still need to be addressed. Phytoremediation is still in its early stage and requires further study to show effective results on the field scale. Based on the achieved data, it is evident, that the plants can be used for phytoremediation of pharmaceuticals and pesticides in line with other already used techniques. It is well known that plants and the accompanying microorganisms play a crucial role in the removal of harmful toxins. Plants can perform phytoremediation only in a certain concentration range, which is different for individual pharmaceutical compounds and pesticides. plants have an excellent potential to be explored in light of the phytoremediation properties they have. The plants uptake contaminants from natural resources like soil and water. Phytoremediation and ecological methods are largely desirable due to their wide acceptance among stakeholders. Several aquatic macrophytes have been reported for the remediation of different categories of wastewater effluents and soil. Among the macrophytes, the common duckweed, L. minor is extensively used for the treatment of pharmaceutical compounds and pesticides. It may be concluded that if the concentration of pharmaceutical compounds and pesticides increases in soil and water, the remediation capacity of plants decreases. The wide application of the plant is due to its ubiquitous nature, invasive mechanism, sporadic reproductive capacity, bioaccumulation potentials, and resilience in a polluted environment. The ultimate objective of such a study is to keep up the ecological balance. The phytoremediation approach is a tried-and-true winner that also represents the direction of remediation techniques. Despite these encouraging efforts so far, the remediation potential still has certain limitations in several areas. To establish practicable remediation strategies in the face of new chemicals that resist conventional remediation techniques, a careful analysis of this method is required. In addition, little research has been done on rhizoremediation and the role of endophytes in the removal of these pesticides. The phytoremediation procedure has the drawback of taking longer to complete than microbial remediation.

References

Abhilash, P. C., & Singh, N. (2010). Effect of growing Sesamum indicum L. on enhanced dissipation of lindane (1, 2, 3, 4, 5, 6-hexachlorocyclohexane) from the soil. *International Journal of Phytoremediation, 12*(5),440-453.

Akinbile, C. O., Ogunrinde, T. A., Che bt Man, H., & Aziz, H. A. (2016). Phytoremediation of domestic wastewaters in free water surface constructed wetlands using Azollapinnata. *International journal of phytoremediation, 18*(1), 54-61.

Allam, A., Tawfik, A., Negm, A., Yoshimura, C., & Fleifle, A. (2015). Treatment of drainage water containing pharmaceuticals using duckweed (Lemnagibba). *Energy Procedia, 74*, 973-980.

Amare, E., Kebede, F., & Mulat, W. (2018). Wastewater treatment by Lemna minor and Azollafiliculoides in tropical semi-arid regions of Ethiopia. *Ecological Engineering, 120*, 464-473.

Anand, S., Bharti, S. K., Dviwedi, N., Barman, S. C., & Kumar, N. (2017). Macrophytes for the reclamation of degraded waterbodies with potential for bioenergy production. In *Phytoremediation potential of bioenergy plants* (pp. 333-351). Springer, Singapore.

Anderson, C. M., & LaBelle, R. P. (2000). Update of comparative occurrence rates for offshore oil spills. *Spill Science & Technology Bulletin, 6*(5-6), 303-321.

Anjali, R., & Shanthakumar, S. (2019). Insights on the current status of occurrence and removal of antibiotics in wastewater by advanced oxidation processes. *Journal of environmental management, 246*, 51-62.

Bedi, J. S., Gill, J. P. S., Aulakh, R. S., & Kaur, P. (2015). Pesticide residues in bovine milk in Punjab, India: spatial variation and risk assessment to human health. *Archives of environmental contamination and toxicology, 69*(2), 230-240.

Bolan, N. S., Park, J. H., Robinson, B., Naidu, R., & Huh, K. Y. (2011). Phytostabilization: a green approach to contaminant containment. *Advances in Agronomy, 112*, 145-204.

Bouldin, J. L., Farris, J. L., Moore, M. T., Smith Jr, S., & Cooper, C. M. (2006). Hydroponic uptake of atrazine and lambda-cyhalothrin in Juncus effusus and Ludwigiapeploides. *Chemosphere, 65*(6), 1049-1057.

Carvalho, F. P. (2017). Pesticides, environment, and food safety. *Food and energy security, 6*(2), 48-60.

Cascone, A., Forni, C., & Migliore, L. (2004). Flumequine uptake and the aquatic duckweed, Lemna minor L. *Water, air, and soil pollution*, *156*, 241-249.

Chaney, R. L., & Baklanov, I. A. (2017). Phytoremediation and phytomining: status and promise. In *Advances in botanical research* (Vol. 83, pp. 189-221). Academic Press.

Chang, G. R. (2018). Persistent organochlorine pesticides in aquatic environments and fishes in Taiwan and their risk assessment. *Environmental Science and Pollution Research*, *25*(8), 7699-7708.

Chaudhry, Q., Schröder, P., Werck-Reichhart, D., Grajek, W., & Marecik, R. (2002). Prospects and limitations of phytoremediation for the removal of persistent pesticides in the environment. *Environmental Science and Pollution Research*, *9*(1), 4-17.

Chen, C., Qian, Y., Chen, Q., Tao, C., Li, C., & Li, Y. (2011). Evaluation of pesticide residues in fruits and vegetables from Xiamen, China. *Food Control*, *22*(7), 1114-1120.

Collins, C. D., Martin, I., & Doucette, W. (2011). Plant uptake of xenobiotics. In *Organic xenobiotics and plants* (pp. 3-16). Springer, Dordrecht.

Cooper, J., & Dobson, H. (2007). The benefits of pesticides to mankind and the environment. *Crop Protection*, *26*(9), 1337-1348.

Datta, R., Das, P., Smith, S., Punamiya, P., Ramanathan, D. M., Reddy, R., & Sarkar, D. (2013). Phytoremediation potential of vetiver grass [Chrysopogonzizanioides (L.)] for tetracycline. *International journal of phytoremediation*, *15*(4), 343-351.

Dickinson, N. M., Baker, A. J., Doronila, A., Laidlaw, S., & Reeves, R. D. (2009). Phytoremediation of inorganics: realism and synergies. *International Journal of Phytoremediation*, *11*(2), 97-114.

Dordio, A. V., Belo, M., Teixeira, D. M., Carvalho, A. P., Dias, C. M. B., Picó, Y., & Pinto, A. P. (2011). Evaluation of carbamazepine uptake and metabolization by Typha spp., a plant with potential use in photo treatment. *Bioresource Technology*, *102*(17), 7827-7834.

Dordio, A. V., Duarte, C., Barreiros, M., Carvalho, A. P., Pinto, A. P., & da Costa, C. T. (2009). Toxicity and removal efficiency of pharmaceutical metabolite clofibric acid by Typha spp.–potential use for phytoremediation? *Bioresource Technology*, *100*(3), 1156-1161.

Ekperusi, A. O., Sikoki, F. D., & Nwachukwu, E. O. (2019). Application of common duckweed (Lemna minor) in phytoremediation of chemicals in the environment: State and future perspective. *Chemosphere*, *223*, 285-309.

El-Queny, F., & Abdel-Megeed, A. (2009). Phytoremediation and detoxification of two organophosphorus pesticide residues in Riyadh Area. *World Applied Sciences Journal*, *6*(4), 570-578.

Etim, E. E. (2012). Phytoremediation and its mechanisms: a review. *Int. J. Environ. Bioenergy*, *2*(3), 120-136.

Fang, L., Liao, X., Jia, B., Shi, L., Kang, L., Zhou, L., & Kong, W. (2020). Recent progress in immunosensors for pesticides. *Biosensors and Bioelectronics*, *164*, 112255.

Fatta-Kassinos, D., Kalavrouziotis, I. K., Koukoulakis, P. H., & Vasquez, M. I. (2011). The risks associated with wastewater reuse and xenobiotics in the agroecological environment. *Science of the Total Environment*, *409*(19), 3555-3563.

Ferro, S., Trentin, A. R., Caffieri, S., & Ghisi, R. (2010). Antibacterial sulfonamides: accumulation and effects in barley plants. *Fresenius Environmental Bulletin*, *19*(9b), 2094-2099.

Forni, C., Cascone, A., Fiori, M., & Migliore, L. (2002). Sulphadimethoxine and Azolla filiculoides Lam.: a model for drug remediation. *Water Research*, *36*(13), 3398-3403.

Fu, F., & Wang, Q. (2011). Removal of heavy metal ions from wastewaters: a review. *Journal of environmental management*, *92*(3), 407-418.

Garbisu, C., & Alkorta, I. (2001). Phytoextraction: a cost-effective plant-based technology for the removal of metals from the environment. *Bioresource Technology*, *77*(3), 229-236.

Garbisu, C., Allica, J. H., Barrutia, O., Alkorta, I., & Becerril, J. M. (2002). Phytoremediation: a technology using green plants to remove contaminants from polluted areas. *Reviews on environmental health*, *17*(3), 173-188.

Garcia-Rodríguez, A., Matamoros, V., Fontàs, C., & Salvadó, V. (2013). The influence of light exposure, water quality and vegetation on the removal of sulfonamides and tetracyclines: a laboratory-scale study. *Chemosphere*, *90*(8), 2297-2302.

Gilden, R. C., Huffling, K., & Sattler, B. (2010). Pesticides and health risks. *Journal of Obstetric, Gynecologic& Neonatal Nursing*, *39*(1), 103-110.

Gujarathi, N. P., Haney, B. J., & Linden, J. C. (2005). Phytoremediation potential of Myriophyllumaquaticum and Pistia stratiotes to modify antibiotic growth promoters, tetracycline, and oxytetracycline, in aqueous wastewater systems. *International journal of phytoremediation*, *7*(2), 99-112.

Halling-Sørensen, B. N. N. S., Nielsen, S. N., Lanzky, P. F., Ingerslev, F., Lützhøft, H. H., & Jørgensen, S. E. (1998). Occurrence, fate, and effects of pharmaceutical substances in the environment-A review. Chemosphere, 36(2), 357-393.

Hoang, T. T. T., Tu, L. T. C., Le, N. P., & Dao, Q. P. (2013). A preliminary study on the phytoremediation of antibiotic contaminated sediment. *International journal of phytoremediation*, *15*(1), 65-76.

Iatrou, E. I., Gatidou, G., Damalas, D., Thomaidis, N. S., & Stasinakis, A. S. (2017). The fate of antimicrobials in duckweed Lemna minor wastewater treatment systems. *Journal of hazardous materials*, *330*, 116-126.

Iori, V., Pietrini, F., & Zacchini, M. (2012). Assessment of ibuprofen tolerance and removal capability in Populus nigra L. by in vitro culture. *Journal of hazardous materials*, *229*, 217-223.

Jesitha, K., & Harikumar, P. S. (2018). Application of nano-phytoremediation technology for soil polluted with pesticide residues and heavy metals. In *Phytoremediation* (pp. 415-439). Springer, Cham.

Kitamura, R. S. A., Brito, J. C. M., Silva de Assis, H. C., & Gomes, M. P. (2022). Physiological responses and phytoremediation capacity of floating and submerged aquatic macrophytes exposed to ciprofloxacin. *Environmental Science and Pollution Research*, 1-18.

Kotyza, J., Soudek, P., Kafka, Z., & Vaněk, T. (2010). Phytoremediation of pharmaceuticals—a preliminary study. *International Journal of Phytoremediation*, *12*(3), 306-316.

Landrigan, P. J., Fuller, R., Acosta, N. J. R., Adeyi, O., Arnold, R., Basu, N., Baldé, A. B., Bertollini, R., Bose-O'Reilly, S., Boufford, J. I., Breysse, P. N., Chiles, T., Mahidol, C., Coll-Seck, A. M., Cropper, M. L., Fobil, J., Fuster, V., Greenstone, M., Haines, A., Zhong, M. (2018). The Lancet Commission on pollution and health. *The Lancet*, *391*(10119), 462-512. https://doi.org/10.1016/S0140-6736(17)32345-0.

Lee, W. Y., Iannucci-Berger, W. A., Eitzer, B. D., White, J. C., & Mattina, M. I. (2003). Plant uptake and translocation of air-borne chlordane and comparison with the soil-to-plant route. *Chemosphere*, *53*(2), 111-121.

Luo, Y.; Guo, W.; Ngo, H.H.; Nghiem, L.D.; Hai, F.I.; Zhang, J.; Ziang, S.; and Wang, X.C. (2014). A review of the occurrence of micropollutants in the aquatic environment and their fate and removal during wastewater treatment. *Science of the Total Environment,* 473-474, 619-641.

Mahmood, Q., Pervez, A., Zeb, B. S., Zaffar, H., Yaqoob, H., Waseem, M., & Afsheen, S. (2013). Natural treatment systems as sustainable ecotechnologies for developing countries. *BioMed research international*, *2013.*

Malik A. (2007). Environmental challenge vis a vis opportunity: the case of water hyacinth. Environ Int., 33(1):122-38. doi: 10.1016/j.envint.2006.08.004. *Epub 2006* Sep 28. PMID: 17010439.

Marsidi, N., Nye, C. K., Abdullah, S. R. S., Hassan, H. A., & Halmi, M. I. E. (2016). Phytoremediation of naproxen in wastewater using vetiver zizaniodes. *J. Eng. Sci. Technol*, *11*(8), 1086-1097.

Matamoros, V., Nguyen, L. X., Arias, C. A., Salvadó, V., & Brix, H. (2012). Evaluation of aquatic plants for removing polar microcontaminants: a microcosm experiment. *Chemosphere*, *88*(10), 1257-1264.

Mercado-Borrayo, B. M., Heydrich, S. C., Pérez, I. R., Quiroz, M. H. & Hill, C. P. D. L. (2015). Organophosphorus and Organochlorine Pesticides Bioaccumulation by Eichhornia crassipes in Irrigation Canals in an Urban Agricultural System, *International Journal of Phytoremediation*, 17:7, 701-708.

Misra, S., & Misra, K. G. (2019). Phytoremediation: an alternative tool towards clean and green environment. In *Sustainable green technologies for environmental management* (pp. 87-109). Springer, Singapore.

Mitton, F. M., Gonzalez, M., Monserrat, J. M., & Miglioranza, K. S. (2016). Potential Use of Edible Crops in The Phytoremediation of Endosulfan Residues in Soil. *Chemosphere*, 148, 300-306. https://doi.org/10.1016/j.chemosphere.2016.01.028.

Morone, A., Mulay, P., & Kamble, S. P. (2019). Removal of pharmaceutical and personal care products from wastewater using advanced materials. *Pharmaceuticals and Personal Care Products: Waste Management and Treatment Technology*, 173-212.

Mukherjee I, & Kumar A (2012) Phytoextraction of endosulfan a remediation technique. *Bull. Environ. Contam. Toxicol.* 88: 250-254.

Ndimele, P. E. (2010). A review on the phytoremediation of petroleum hydrocarbon. *Pakistan journal of biological sciences: PJBS*, *13*(15), 715-722.

Nwaichi, E. O., Frac, M., Nwoha, P. A., & Eragbor, P. (2015). Enhanced phytoremediation of crude oil-polluted soil by four plant species: effect of inorganic and organic bioaugmentation. *International journal of phytoremediation*, *17*(12), 1253-1261.

Nwaichi, E. O., Osuji, L. C., & Onyeike, E. N. (2011). Evaluation and decontamination of crude oil-polluted soils using CentrosemapubescenBenth and amendment-support options. *International journal of phytoremediation*, *13*(4), 373-382.

Obot, E., Antonio, Q. B., Braide, S., Dore, M., Wicks, C., & Steiner, R. (2006). Niger delta natural resource damage assessment and restoration project. *Rep. Submit. to Fed. Minist. Environ. Abuja Niger. Conserv. Found. Lagos.*

Okpokwasili, G. C., & Odokuma, L. O. (1990). Effect of salinity on biodegradation of oil spill dispersants. *Waste management*, *10*(2), 141-146.

Otani T, Seike N, Sakata Y. 2007. Differential uptake of dieldrin and endrin from soil by several plant families and Cucurbita genera. Soil Sci Plant Nutr, 53, 86– 94.

Pereira, R. C., Camps-Arbestain, M., Garrido, B. R., Macías, F., & Monterroso, C. (2006). The behavior of α-, β-, γ-, and δ-hexachlorocyclohexane in the soil–plant system of a contaminated site. *Environmental Pollution*, *144*(1), 210-217.

Petrovic, N., Walden, T. B., Shabalina, I. G., Timmons, J. A., Cannon, B., & Nedergaard, J. (2010). Chronic peroxisome proliferator-activated receptor gamma (PPARgamma) activation of epididymally derived white adipocyte cultures reveals a population of thermogenically competent, UCP1-containing adipocytes molecularly distinct from classic brown adipocytes. *The Journal of biological chemistry*, *285*(10), 7153–7164. https://doi.org/10.1074/jbc.M109.053942.

Petrovic, P., Dietrich, T., Fransson, P., Andersson, J., Carlsson, K., & Ingvar, M. (2005). Placebo in Emotional Processing— Induced Expectations of Anxiety Relief Activate a Generalized Modulatory Network. *Neuron,* 46(6), 957-969. https://doi.org/10. 1016/j.neuron.2005.05.023.

Proia, L., Lupini, G., Osorio, V., Pérez, S., Barceló, D., Schwartz, T., Amalfitano, S., Fazi, S., Romaní, A., & Sabater, S. (2013). Response of biofilm bacterial communities to antibiotic pollutants in a Mediterranean river. *Chemosphere,* 92(9), 1126-1135. https://doi.org/10.1016/j.chemosphere.2013.01.063.

Reinhold, D., Vishwanathan, S., Park, J. J., Oh, D., & Saunders, F. M. (2010). Assessment of plant-driven removal of emerging organic pollutants by duckweed. *Chemosphere*, *80*(7), 687-692.

Riaz, N., Havel, J. J., Makarov, V., Desrichard, A., Urba, W. J., Sims, J. S., Hodi, F. S., Martín-Algarra, S., Mandal, R., Sharfman, W. H., Bhatia, S., Hwu, W. J., Gajewski, T. F., Slingluff, C. L., Jr, Chowell, D., Kendall, S. M., Chang, H., Shah, R., Kuo, F., Morris, L. G. T., Chan, T. A. (2017). Tumor and Microenvironment Evolution during Immunotherapy with Nivolumab. *Cell*, *171*(4), 934–949. e16. https://doi.org/10.1016/j.cell.2017.09.028.

Rissato, S. R., Galhiane, M. S., Fernandes, J. R., Gerenutti, M., Gomes, H. M., Ribeiro, R., & de Almeida, M. V. (2015). Evaluation of Ricinus communis L. for the Phytoremediation of Polluted Soil with Organochlorine Pesticides. *Biomed Res Int.,* 2015:549863. doi: 10.1155/2015/549863. Epub 2015 Aug 2. PMID: 26301249; PMCID: PMC4537713.

Romeh, A. A. (2015). Evaluation of the phytoremediation potential of three plant species for azoxystrobin-contaminated soil. *Int. J. Environ. Sci. Technol.* 12, 3509–3518. https://doi.org/10.1007/s13762-015-0772-7.

Ryšlavá, H., Pomeislová, A., Pšondrová, Š., Hýsková, V., & Smrček, S. (2015). Phytoremediation of carbamazepine and its metabolite 10, 11-epoxycarbamazepine by C3 and C4 plants. *Environmental Science and Pollution Research*, *22*(24), 20271-20282.

Schaefer, M., & Juliane, F. (2007). The influence of earthworms and organic additives on the biodegradation of oil-contaminated soil. *Applied soil ecology*, *36*(1), 53-62.

Suresh B, Sherkhane P D, Kale S, Eapen S, Ravishankar G A (2005) Uptake and degradation of DDT by hairy root cultures of Cichorium intybus and Brassica juncea. *Chemosphere*, 61(9):1288–1292.

Susarla, S., Medina, V. F., & McCutcheon, S. C. (2002). Phytoremediation: an ecological solution to organic chemical contamination. *Ecological Engineering*, *18*(5), 647-658.

Truua J., Truua, M., Espenberga, M., Nolvaka, H., & Juhanson, J. (2015). Phytoremediation and plant-assisted bioremediation in soil and treatment wetlands. *A review. Open Biotechnol.* J., 9: 85–92.

Udeh, N. U., Nwaogazie, I. L., & Momoh, Y. (2013). Bioremediation of crude oil-contaminated soil using water hyacinth (Eichhorniacrassipes). *Advances in Applied Science Research*, *4*(2), 362-369.

USDA U.S. Department of Agriculture), 2018. *The PLANTS Database, National Plant Data Team.* NRCS, United States Department of Agriculture, Greensboro, NC 27401-4901 USA. http://plants.usda. Gov.

Van De Steene, J. C., Stove, C. P., & Lambert, W. E. (2010). A field study on 8 pharmaceuticals and 1 pesticide in Belgium: Removal rates in waste water treatment plants and occurrence in surface water. *Science of The Total Environment,* 408(16), 3448-3453. https://doi.org/10.1016/j.scitotenv.2010.04.037.

Van der Werf, H. M. (1996). Assessing the impact of pesticides on the environment. *Agriculture, Ecosystems & Environment*, 60(2-3), 81-96. https://doi.org/10.1016/S0167-8809(96)01096-1.

Wojcieszyńska, D., Domaradzka, D., Hupert-Kocurek, K., & Guzik, U. (2014). Bacterial degradation of naproxen–an undisclosed pollutant in the environment. *Journal of environmental management*, *145*, 157-161. https://doi.org/10.1016/j.jenvman.2014.06.023.

Wu, S, Zhang, L., & Chen, J. (2012). Paracetamol in the environment and its degradation by microorganisms. *Applied microbiology and biotechnology*, *96*(4), 875-884.

Zhang, D. Q., Hua, T., Gersberg, R. M., Zhu, J., Ng, W. J., & Tan, S. K. (2012). The fate of diclofenac in wetland mesocosms planted with Scirpusvalidus. *Ecological Engineering*, *49*, 59-64.

Zhou, H., Liu, X., Chen, X., Ying, T., & Ying, Z. (2018). Characteristics of removal of waste-water marking pharmaceuticals with typical hydrophytes in the urban rivers. *Science of the Total Environment*, *636*, 1291-1302.

Chapter 8

Wetlands Classification, Importance, Causes of Degradation and Ecological Methods of Restoration

Sonam Gupta*
and Aneet Kumar Yadav
Department of Environmental Science, Babasaheb Bhimrao Ambedkar University, Vidya Vihar, Raebareli Road, Lucknow, Uttar Pradesh, India

Abstract

Wetlands have been crucial to the development of human civilizations and cultural history. Wetlands are one of the planet's most functional ecosystems, offering a variety of biological, ecological, economic, social, and cultural services that contribute to human welfare. Despite their indispensable values, wetlands are under tremendous pressure and are declining in their ecological, hydrological, and economic functions parallel shrinking in aerial extent. However, the continuous progression of global degradation of wetlands is also due to a lack of information on their importance and methods of restoration. This chapter focuses on the classification system for wetlands and highlights the importance of wetlands from ecological and economic perspectives. This chapter also focuses on growing threats due to urbanization, agricultural practices, livestock grazing, and pollution. In the chapter's final section, restoration methods are discussed to reverse and protect these delicate ecosystems.

Keywords: wetland, degradation, ecological restoration, pollution

---

* Corresponding Author's Email: guptasonam3333@gmail.com.

In: Conserving, Halting and Reinstating Degraded Natural Resources …
Editors: Rajani Srivastava and Pankaj Kumar
ISBN: 979-8-89113-067-8

1. Introduction

Wetlands are the transitional zone between terrestrial and aquatic ecosystems that comprise marsh, swamps, fen, peatlands, ponds, lakes, waterlogged land, and ash pond (Mahdavi, 2018). They are the most abundant and biologically rich ecosystems on the earth and play a significant role in the ecological sustainability of the region.

Wetlands play myriad roles such as improving water quality, climate change, biodiversity conservation, sediment trapping, water purification, ecotourism, and mitigation of climate change, flood, and soil erosion. Wetlands have been dubbed the "kidneys of Earth" due to their decontamination abilities such as sedimentation, filtration, adsorption, and biological assimilation. They cover about 4–6% area which accounted for 40.6% of global ecosystem services (Dworrak et al., 2022). Thus, directly or indirectly wetlands helped millions of people through their physical, biological, social, cultural, and economic value and provide habitat and support a diverse range of biodiversity. Wetland ecosystem services have the highest value per hectare of any type of ecosystem, and their total values account for 47% of global ecosystem values. It is one of the most important and productive ecosystems on the planet (Xu et al., 2019). The summary of the major ecosystem services provided by wetlands is given in Table 1.

Wetlands support up to 40% of the world's species, but more than 25% of all wetland plants and animals are endangered (Satyaveer et al., 2022). Wetlands are one of the most endangered ecosystems in the world, depleting more than three times faster than forest ecosystems. Since the 1700s, nearly 90% of the world's wetlands have been degraded and since 1970 (50 years) 35% of world wetlands have been degraded due to several anthropogenic activities such as overexploitation of natural resources, climate change, increase pollution levels, construction, overfishing, and others (Janse et al., 2019). Solutions are critical for the long-term survival of wetlands, particularly in closely populated areas (Davenport et al., 2010).

The Ramsar Convention's primary goal is to prevent global wetlands degradation through wise use and sustainable management. The Ramsar Convention promotes the designation of sites with a representative, uncommon in addition to wetlands crucial to the preservation of biodiversity. The convention requires that wetland encroachment and loss be limited now and in the future (Davidson, 2018). It is one of the most important treaties for advancing the ecosystem approach to conservation, and it protects the entire habitat as a single unit and system (Griffin & Ali, 2014). Various wetlands restoration methods such as topographical excavation, soil development and transplantation, plantation, flow regulation, and recovery of other ecosystem services have gained popularity over the years.

The natural attributes of the site need to be restored for the successful restoration of other aspects like quality of water and restoration of the native flora and fauna. It can be achieved through processes like stream channelization, ditching in wetlands, disconnection from the adjacent ecosystem, shoreline modification, and so on. For a successful restoration of the wetland natural functions in the site should be identified, and the list should be prioritized as per the importance of the functions and the functions to be restored in the degraded wetland system.

With this frame of reference, the objective of this chapter is to analyze the existing classification systems of wetlands, the importance of wetlands, threats to wetlands, and different methods of restoration.

Table 1. Major ecosystem services and their functions

Ecosystem Services	Functions	References
Nutrient Cycling	They play a vital role in primary productivity and nutrient cycling. The wetlands' nutrient profile is in the form of input of sewage and agricultural runoff, carrying a lot of pesticides and fertilizers. Wetlands macrophytes offer nutrition for waterfowl and other living creatures of the wetland. The roots of the aquatic plants hold the sediment in place and prevent it from being carried off.	Johnes et al., (2020)
Soil formation (Supporting)	Wetlands support soil formation by sediment retention and accumulation of organic matter. Wetlands also support the storage, recycling, processing, and acquisition of nutrients.	Osland et al., (2018)
Pollination	Wetlands also help to protect pollinators and other beneficial insects. Pollinator habitat requirements, the importance of native plants in wetland communities, and how to determine which plant species are appropriate for your region to increase biodiversity and improve wildlife habitat.	Qiu (2019)
Climate regulation	Wetlands act as a source and sink for greenhouse gases and influence local and regional temperature, precipitation, and other climatic processes.	Tan et al., (2020)
Food Production	Wetlands can be converted to increase food production because arable land is limited. The development of wetlands promotes social and economic development in industries such as agriculture, fisheries, and aquaculture.	Sulaiman et al., (2019)

Table 1. (Continued)

Ecosystem Services	Functions	References
Cultural	Many religions attach spiritual and religious values to the aspect of wetland ecosystems. Wetlands also provide opportunities for formal and informal education and training.	Pueyo-Ros et al., (2019)
Groundwater Recharge	Wetlands are one of the most productive ecosystems and play an important role in the hydrological cycle. It supports the hydrological flows by recharging and discharging the groundwater.	Abraham et al., (2022)
Detoxification	Wetlands help in water purification and waste treatment by retention, recovery, and removal of excess nutrients and other pollutants. Wetlands are bioengineers of the ecosystem used for edge treatments, flood mitigation, and water treatment, etc.	Kumar et al., (2021)
Wastewater treatment	Wetland plants may adsorb and accumulate pollutants from wastewater in their tissue as well as provide a complimentary increasing environment for microorganisms in the wetlands system.	Almuktar et al., (2018)
Biodiversity	Wetland supports rich and diverse varieties of flora and fauna. Different varieties of bird habitats in wetland ecosystems include migratory birds as well as IUCN red-listed species. Rich aquatic life including the phyto and zooplanktons, crabs, fishes, aquatic plants, and mangroves survive in wetlands. Wetland ecosystem also supports different types of plant species which supports the ecosystem functioning.	Xu et al., (2019)
Recreational	Wetlands are a source of income for local people by utilizing wetland services. Among them, fishing and hunting are the most important ways; others include bird watching and boating. The concept of ecotourism and the recreational values of wetlands has been widely used by the local people to generate their income and these policies are also supportive of achieving the goals of sustainable development at the local and regional levels.	Osland et al., (2018)

2. Wetland Classification and Types

The most widely used definition of a wetland is given by Cowardin (1979), which states that a wetland must satisfy at least one of the following conditions: it should occasionally support hydrophytes or substrate is non-soil and is periodically saturated with water or submerged beneath shallow water or the substrate is formed of undrained hydric soil. Several wetland categorization systems have adopted this specific criterion as their foundation for classification.

One of the most complete and extensively used classification schemes for wetland types was created by Cowardin & Golet (1995). This hierarchical structure can be employed in conjunction with other hydrological, soil, and chemical modifiers. This serves as the foundation for several classification schemes such as Alberta Environment Sustainable Resource Development 2015, and US Federal Geographic Data Committee (2013). Palustrine wetlands are categorized into seven different groups based on the composition of the substrate and size of the dominant life form. The rationale behind choosing life form and substrate as classification criteria was that they would be simple to identify, even on satellite photography, and that they do not alter with the seasons.

The Ramsar "Classification System for Wetland Type" is a comprehensive framework for the rapid identification of wetlands, and it was adopted at the Conference of the Contracting Parties in 1990 with subsequent amendments. It classifies wetlands into 3 major categories and 42 kinds of wetlands. This classification is used for rapid assessment and uses a range of randomly selected distinguishing factors such as substrate type, hydroperiod, geomorphic shape, water quality, and vegetation type (Arya et al., 2020).

To provide a more precise evaluation Brinson (1993) and Semeniuk & Semeniuk (1995) primarily focused on hydrogeology and landforms for assessing the physical, chemical, and biological functions of different types of wetlands. They emphasized hydrogeomorphic processes to categorize wetlands across a variety of climatic, geological, soil, and vegetation environments. Key prediction is based on hydrogeology and landforms because these are thought to be major factors influencing wetland development in the geomorphic classification scheme used for Australian inland wetlands. For example, in a dryland setting, a flat might be constantly dry, whereas, in a humid environment, a flat might be a wetland.

A thorough process-based classification of floodplains is provided by Nanson and Croke (1992), and it is based on a river's capacity to entrain and carry sediment along with the floodplain's resistance to erosion. The system is split into three divisions based on the major mechanisms that produce floodplains: low-energy cohesive, medium-energy non-cohesive, and high-energy non-cohesive. As a result, the ensuing floodplains are unique in their design and origin. Additionally, the classification acknowledges that floodplains can alter from one kind to another because it is dependent on processes.

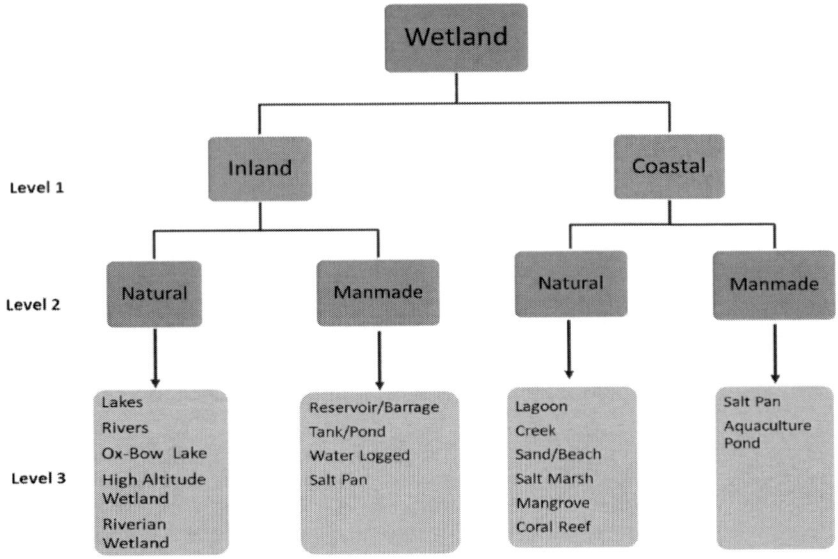

Source: SAC, 2011.

Figure 1. Classification of Indian wetlands.

A categorization system for all inland aquatic ecosystems in South Africa, including rivers and wetlands, was developed by Ewart-Smith et al., (2006) and Ollis et al., (2015). The wetland categorization builds on earlier work by Kotze et al., (2009), with the subsequent studies mostly based on Brinson (1993). This approach is similarly hierarchical and includes the landscape at three different levels, such as the valley bottom, bench, plain, or slope. Hydrogeomorphic units are added at level four. These units, which include floodplains, rivers, depressions, seeps, channeled and unchanneled valley bottoms, and wetland flats, are defined by landform, hydrological feature, and hydrodynamics.

In India, wetlands can be found from the Himalayan region to the Deccan Knap. A wide variety of climatic and geographical conditions exist in India, which contribute to the country's abundant flora and wildlife (Prasad et al., 2002). Indian wetlands were divided into three levels in the National Wetlands Atlas of 2011, spanning from Level 1 to Level 3 (Figure 1). Inland and coastal locations make up the two classes of Level-1 wetlands, while man-made and natural wetlands make up the two classes of Level-2 types. However, both natural and artificial wetlands fall under the Level 3 category, and these are further classified into 19 subgroups based on topographical, climatic, and temporal factors.

3. Importance of Wetlands

Wetlands' unique ecological features provide several products and services to humans. Some of the major services are discussed below:

3.1. Multiple Water Services

Wetlands have been providing multiple water services such as water for domestic use, irrigation, fisheries, and recreational purposes for decades. Wetlands also play a significant role in groundwater recharge, flood control, and silt collection. In addition to irrigation, tanks are used for storing water for domestic purposes, cultivation, gathering fodder grass, and making bricks W (Kumar et al., 2013).

Furthermore, from an ecological standpoint, tanks are important as they preserve biodiversity, soil, and water (Kumar & Vedantam, 2016). Additionally, tanks help recharge the groundwater, manage floods, and catch silt (Kumar et al., 2012). Many lakes like Dal Jheel (Jammu and Kashmir), Chilka (Orissa), Kolleru (Andhra Pradesh), Carambolim (Goa), Nainital (Uttarakhand), and Vembanad (Kerela) have long provided services for recreation, tourism, fisheries (Jain et al., 2007).

A rich and diversified array of aquatic vegetation and fauna are supported by these lakes, which also aid in groundwater recharging. Wetlands have a big impact on fish production in India. Currently, inland water bodies including rivers, canals, reservoirs, tanks, ponds, and lakes produce the majority of the nation's fish (61% of all production) (Bassi et al., 2014). About 80% of the production of inland aquaculture is carp. Andhra Pradesh now leads in terms of total inland fish production followed by West Bengal, Bihar, and Uttar Pradesh (Department of Fisheries, 2020). Overall fisheries contribute 1.2% of India's GDP and 5.4% of the agricultural GDP (Agrawal et al., 2016).

3.2. Carbon Sequestration

The carbon cycle is significantly influenced by marshes, swamps, mangroves, and peatlands. Wetland biomass and dissolved elements in the surface and groundwater are the short-term carbon storage sources in wetlands, whereas sediments are the long-term source (Mu et al., 2013). Wetlands' contribution to methane emissions is estimated to be around 40% globally yet they have the highest carbon density of all terrestrial ecosystems and possess a comparatively higher capacity to sequester carbon dioxide. They tend to absorb carbon dioxide due to high rates of organic inputs and low rates of decomposition (Pant et al., 2003).

There could be 200 times more carbon in wetlands' soils than in their vegetation. However, extensive wetland drainage and subsequent widespread cultivation had turned them into a net producer of CO_2. By being restored, wetlands can become a sink for CO_2 in the atmosphere (Lal, 2008). According to estimates, the carbon sequestration capability of restored wetlands (during 50 years) equals roughly 0.4 tonnes C/ha/year (Chandra & Kumar, 2020). Mangroves can trap roughly 1.5 metric tonnes of carbon per hectare each year (Kathiresan & Thakur, 2008). Depending on their bio-geochemical processes and hydrology, wetlands either serve as net gas producers or sequesters. To determine whether wetlands can be maintained as long-term net carbon sinks and their potential contribution to the global carbon trading system and climate change mitigation, more research is therefore needed.

3.3. Pollution Abatement

In many agricultural and urban environments, wetlands serve as a sink for toxins. It acts as a low-cost technique for mitigating point and non-point pollution from an economic standpoint (Bystrom et al., 2000). Riparian wetlands absorb phosphate and nitrate from the surface and sub-surface runoff, thus reducing the nutrient load of through-flowing water (Verhoeven et al., 2006). The range of the maximum annual rates of nitrogen and phosphorus removal by wetlands in temperate climates is 1000–3000 kg N/ha and 60–100 kg P/ha, respectively (Groffman & Crawford, 2003). However, as a result of increasing urbanization and changes in land use that harm water quality and biodiversity, alter nitrogen cycling, and cause a loss of species, wetlands are being loaded with nutrients at a rate that far surpasses their ability to hold contaminants and eliminate them through sedimentation, nitrification, adsorption, and accumulation by aquatic plants (Verhoeven et al., 2006).

The removal of heavy metals from wetlands has been demonstrated by several plants, including *Typha, Phragmites, Eichhornia, Azolla*, and *Lemna*

(Rai, 2008). Constructed wetlands are regarded as a practical method for treating sewage from urban areas. However, properly designed wetlands that have been restored or created can be used as pollution sinks (Al Hadidi, 2021).

3.4. Flood Control

Wetlands are crucial for flood prevention. Wetlands absorb water and slow the flow of floodwater, which helps to decrease the effects of floods. Additionally, they catch nutrient load and suspended particles during times of flooding. Therefore, streams that enter rivers through wetlands will transfer fewer nutrients and suspended particles to the rivers than streams that enter rivers directly. Wetlands are considered a natural capital equivalent for traditional flood control investments such as dykes, dams, and embankments because of their effectiveness related to preventing flood damage (Boyd & Banzhaf, 2007).

Furthermore, the reduction in wetland areas due to conversion to agricultural purposes, like rice growing and fish pond aquaculture has aggravated flood-induced disasters in the last decade (Prasad et al., 2002). Additionally, as wetlands receive inflows from shallow aquifers as well, increased groundwater pumping for agriculture in eastern India had negatively impacted them. Shallow wetlands may temporarily dry up because of the shallow aquifers' dropping water tables throughout the winter and summer months when agricultural water demand increases (Kumar et al., 2012)

3.5. Biodiversity Hotspot

Wetlands play a crucial role in fostering species variety, just like any other natural ecosystem. While certain vertebrates and invertebrates depend on wetlands for the entirety of their lives, others only associate with them at certain times. Wetlands play a vital role in the maintenance of food chains because they offer a setting for photosynthesis and the recycling of nutrients (Kattel et al., 2020). Western ghats in southern India are known for their flora and fauna diversity. Wetlands are crucial places for animal reproduction and offer migratory birds a haven.

Many migratory bird species from western and European countries visit India's wetland regions during the winter, including Bharatpur Wild Life Sanctuary in Rajasthan, Little Rann of Kutch, and coastal portions of Saurashtra in Gujarat. According to certain estimates, India is home to between 1200 and 1300 species of migratory birds, or around 24% of the country's total bird species (Agarwal, 2011). Additionally, efforts have been undertaken to value wetland biodiversity. In various sites along the Elbe River in Germany, the value of enhancing biodiversity through artificial wetlands is approximately USD 1942 per hectare per year (Ghermandi et al., 2008). Similarly, the value of the tropical river and inland fisheries alone has been estimated at USD 5.58 billion annually (Neiland, 2008). In India, about 20 million people rely on inland fisheries and livelihood, this presents a major opportunity (Das et al., 2022).

4. Causes of Degradation

Wetlands are one of the most versatile ecosystems on earth, serving a variety of social, cultural, economic, biological, and ecological needs for humans (Arya et al., 2020). Despite their contributions, the majority of the world's wetlands are degrading at an alarming rate. Wetlands are dynamic ecosystems that undergo continuous natural change as a result of siltation, erosion, subsidence, and drought. Human activities on land and in water are putting wetlands under intense pressure and endangering the natural functions they offer. The vital functions, values, and characteristics of wetlands are significantly impacted by direct and indirect anthropogenic interventions, even though natural change is typical and expected. Degradation of wetland ecosystems could result in loss of value and function. The following are the major causes of wetland degradation:

4.1. Improper Agriculture Practices and Expansions

Underlying causes putting pressure on wetlands include a lack of available agricultural land brought on by rising human and livestock populations and limited community understanding of the ecological advantages of wetlands (Reid et al., 2019). Soil erosion, loss of the continental shelf, and decreased soil fertility are all consequences of inefficient farming practices. Thus, farmers become both perpetrators and victims of environmental degradation (Yitaferu et al., 2004) as a result of low agricultural output. Wondie (2012) states that due to upstream intense agricultural operations and deforestation, wetlands in the Lake Tana basin are currently in danger from sedimentation (silt load). Compared to point sources of pollution, non-point sources, such as agricultural runoff, have more negative consequences. In the last two decades, farming has moved to wetlands such as river banks, channel banks, and lake shorelines due to increased population pressure and resource limitations (land, water) (Leykun, 2003).

4.2. Urbanization and Industrialization

Wetland recreational qualities have been significantly impacted by expanding urbanization, industry, and unplanned tourism development operations along lake and river banks. Wetlands have suffered negative effects from human activities like agriculture, grazing, brick building, settlement, and gathering of grass and reeds (Mengesha, 2017). The degradation of wetlands is intimately related to the growth of urban areas across the entire nation. Wetlands that are close to urban areas are deteriorating as a result of growing industry sources such as hotels, health centers, and factories. For instance, the quantity of solid and liquid wastes produced by various sources is growing in volume and variety. Along with urbanization, industries in the wetlands have a significant negative impact on their regular operation, making it more difficult for them to filter wastewater and prevent stream siltation (Boyd, 2020).

4.3. Livestock Grazing

In addition to the aforementioned issues, grazing has been identified as a threat to wetlands. Wetlands lose their functional characteristics when continual cultivation is followed by grazing primarily by livestock leading to soil compaction which results in the loss of vegetation (Morris & Reich, 2013).

Livestock kills natural flora and compacts and tramples the soil. They enhance water flow and degrade drainage channels that lead to gullies (Avila et al., 2021). Because of the decreased water table and altered vegetation, these factors frequently cause wetlands to completely degrade. To maintain the integrity of the systems that sustain life and promote sustainable socio-economic development, wetlands generate an ecological balance. However, a lot of wetland ecosystems, especially flood plains, and swamps, are considered wastelands and are being destroyed at an alarming rate (Giweta & Worku 2018).

4.4. Week Institutional Capacity

Threats to wetlands also increase due to limited knowledge of wetlands' current status, their values, and lack of ability to put government plans into practice (Alikhani et al., 2021). Even the need for their conservation and sustainable use is not widely understood. The issues previously described are a result of a lack of clear understanding among the general public, decision-makers, and policymakers, as well as a lack of clear policy and direction on the wetlands issue. Wetland resources are currently being exploited more than they can regenerate. It appears that there is no clear understanding of who truly has control over such precious resources in terms of the arrangement of land tenure. The situation compels us to ponder whether wetlands are deteriorating due to policy inadequacies or legal frameworks.

4.5. Deforestation and Degradation of Catchments

Deforestation, excessive grazing, unsustainable farming methods, and wetland degradation are all causes of land degradation. As a result of population development and the ensuing increase in farming, wetlands are allegedly constantly in danger of being destroyed by deforestation (Kingsford, 2016). The need for fuel and wood building is growing. Sediment loads and the conversion, eradication, and invasion of significant natural buffers like wetlands limit the lakes' capacity to withstand stress. It is reported that 65% of wetland disturbances have human origins, with the remaining 35% having natural origins. 73% of these disruptions are assumed to have direct human causes, whereas the remaining 27% are thought to have indirect causes (Gebresllassie et al., 2014). Increased human intrusion leads to enhanced harvesting and burning of emergent macrophytes with continued growth in submerged macrophytes causing imbalance.

4.6. Pollution

Inappropriate human behavior is a major cause of pollution. The main pollution threats to sustainably managed wetland resources are poor chemical management and inadequate waste disposal methods. Agricultural runoff of pesticides and fertilizers and discharge of untreated municipal and industrial water pollutes lakes and streams stimulating algal growth. Pollution of lakes also reduces or eliminates fish populations and leads to the loss of many cultural services provided by them (Bassi et al., 2014). Pollution affects the overall ecological functioning of the wetland environment.

4.7. Overexploitation of Wetland Resources

The poorest households are primarily dependent on the wetland because they have no other means of generating income. Another serious hazard to this ecosystem is the over-harvesting of wetlands resources by residents, since most portion of the wetland is managed by the community. Other significant hazards include excessive water extraction from wetlands, erosion, and sedimentation.

5. Ecological Methods for Restoration of Wetland

Wetland restoration methods are classified into three types based on methodology: active restoration, passive restoration, and creation (Figure 2) (Mitsch & Gosselink, 2007).

5.1. Active Restoration

Active restoration approaches are based on physical intervention in which human involvement increases the restoration process. The onsite analyses with other sources like surveys, interviews, and literature studies are referred to as the issues faced by a wetland ecosystem. The studies explain that the restoration of a wetland ecosystem can be achieved by restoring the abiotic conditions that support and emphasize biotic succession. Along with these the practices that prevailed in the area also have to be restored to support ecological restoration.

Active methods of restoration include re-contouring the site to provide the proper substrate for native species, changing water flow in water structures (such as weirs or culverts), intensive planting, seeding, intensive non-native species control, and bringing soils to the site to provide the right substrate for native species, such work can have a significant design, engineering, construction, and costs (Zhao et al., 2016).

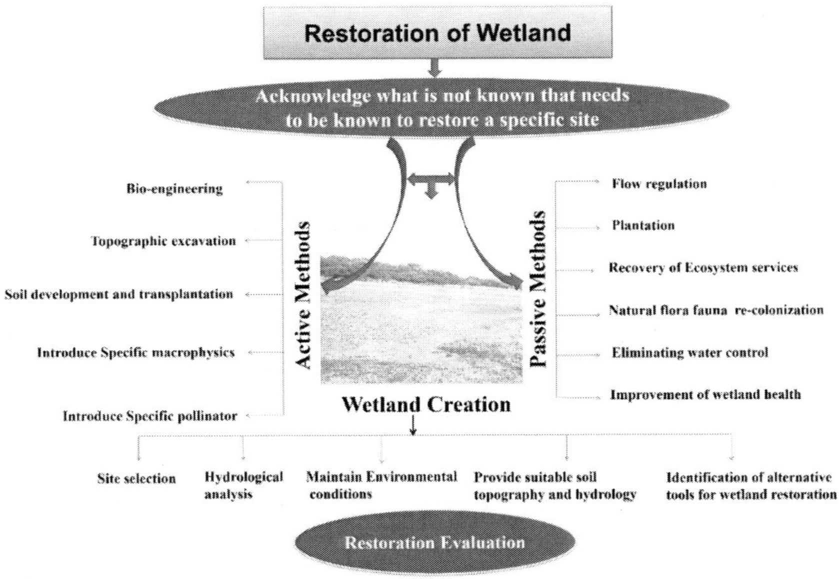

Figure 2. Methods of wetland restoration.

Humans must maintain control and regularly intervenes to restore, recreate, or improve the community structure and ecosystem processes in wetlands (Wagner et al., 2008). Wetland topography is frequently altered, water flow is changed by water management structures (such as division dikes), and the soil is moved, when plants are planted (Zhang *& Yuqin.,* 2009). The restoration measures included the construction of dams and embankments, the collection of freshwaters, and the diversion of water.

5.2. Passive Restoration

Passive restoration is dependent on natural processes. It deals with minimum alterations at the site. For example, for a wetland with altered hydrology, restoring the original hydrological regime may be enough for the whole site to re-establish the whole ecological integrity.

Passive restoration methods entail removing factors that contribute to the degradation of coastal wetlands and restoring degraded wetlands to natural conditions in healthy states (Wagner et al., 2008). Passive methods focus on improving eco-hydrological processes to rebuild the hydrogeomorphology for wetland self-restoration (Jarzemsky et al., 2013). Converting cropland to marshes and grasslands can help to mitigate the degradation of marshes caused by canalization (Zhao et al., 2016).

5.3. Creation

The process of converting lands that previously had no wetlands into healthy wetlands is referred to as creation (Broome et al., 2019). Wetland development necessitates local conditions, particularly appropriate hydrology. Providing water where water is not available naturally and establishing natural vegetation on non-hydric soils is the challenging step in wetlands creation. While it is possible to create, it usually takes far more planning and effort than restoration projects and the results are unpredictable. The natural wetlands provide only a subset of wetland functions and many attempts to convert upland to wetland. Wetlands can more easily be created in open water, but this process frequently involves filling in already-existing aquatic habitats with dirt and other materials that can destroy aquatic life. The results of the restoration may depend on the level of degradation and other environmental factors. Creation or enhancement may be the best choice in some circumstances, but restoration is typically more likely to result in better wetland resources (Zhao et al., 2016).

5.4. Other Wetland Restoration Techniques

The techniques used to restore or create wetlands vary depending on the type of wetlands involved. Salt marsh restoration can be accomplished through a variety of methods, including hydrology restoration, weed management, plant intrusion control, reducing vertebrate disturbance, grazing regime enforcement, and pollution control (Adam et al., 2010). Techniques such as digging and tillage, cutting and waterlogging, mowing, breaking of rhizomes, and biological substitution with *Phragmites australis* proved effective in mitigating the degradation of wetlands. Recent studies have explicitly considered wetlands that have been restored or created from a multipurpose standpoint.

In brief, it could be summarized that wetland restoration methods and techniques have improved and restoration policies have got better with time and experience. Some degradation is reversible, but many damages are irreversible. However, since each project is unique creating a restoration pattern is difficult. Therefore, adaptive restoration holds great promise to restore specific sites (Elliott et al., 2007).

Conclusion

Wetland ecosystems are found throughout a variety of topographic and climatic regimes and support a variety of distinctive habitats. They are regarded as an essential component of the hydrological cycle and are extremely productive systems in their undisturbed states. Wetlands provide a variety of ecological services such as water for irrigation, domestic water supply, freshwater fisheries, and recreation. Additionally, they serve a crucial role in the replenishment of groundwater, carbon sequestration, flood prevention, and reduction of pollutants. Wetlands are among the highly threatened ecosystem and are being lost and degraded at a much faster rate than any other ecosystem affecting the entire world. Numerous wetlands are impacted by anthropogenic pressures in rural and urban areas such as pollution from industries and households, tourism, land use change in watersheds, encroachments, and overexploitation of natural resources.

The previously described issues are exacerbated by the general public's and decision and policy makers' lack of awareness, as well as by the lack of a clear policy and direction on the wetlands issue. To support healthy riparian, aquatic, and wetland habitats, water quality must be maintained and preserved. The biological, physical, and chemical integrity of the system must be preserved, and the survival, growth, migration, and reproduction of the creatures that make up aquatic and riparian ecosystems must be supported. As part of comprehensive land management and watershed management, local governments and communities with an interest in wetlands should adopt wetland preservation and adaptation goals and policies. Moreover, the majority of the studies on wetland focus on the limnological elements and ecological economics of wetland management. Yet, the physical and socio-economic factors causing limnology have not been thoroughly investigated. Furthermore, the institutional features of wetland management (policies, norms, regulations, and organizations) have only recently captured the attention of researchers. To develop more comprehensive management strategies for wetlands that are threatened by several climatic and anthropogenic factors, more research including the physical, socio-economic, and institutional factors influencing the conditions of wetlands and their use is therefore necessary.

References

Abraham, T., Muluneh, A., Girma, R., Hartmann, A., & Tekleab, S. (2022). Quantifying sensitivity of groundwater recharge to land use and land cover changes by improving model performance on the wetland dominated Tikur Wuha watershed, Ethiopia. *Water Cycle,* 3 112–125. https://doi.org/10.1016/j.watcyc.2022.08.002.

Adam, E., Mutanga, O., & Rugege, D. (2010). Multispectral and hyperspectral remote sensing for identification and mapping of wetland vegetation: A review. *Wetlands Ecology and Management,* 18(3), 281–296. https://doi.org/10.1007/s11273-009-9169-z.

Agarwal, M. (2011). *Migratory birds in India: Migratory birds dwindling New Global Indian* (http://newglobalindian.com/nature).

Agrawal, R., Darapuneni, R. R., V L N Rao, Banda V. L. N., Nanda, S. K., & Bhattacharya, S. (2016). Assessment of employment potential for fisheries professionals in India. *Indian Journal of Fisheries,* 63(2), 102–109.

Al Hadidi, L. A. (2021). Constructed wetlands a comprehensive review. International *Journal of Research -GRANTHAALAYAH,* 9(8), 395–417. https://doi.org/10.29121/granthaalayah.v9.i8.2021.4176.

Alikhani, S., Nummi, P., & Ojala, A. (2021). Urban wetlands: A review on ecological and cultural values. *Water,* 13(22), 3301. https://doi.org/10.3390/w13223301.

Almuktar, S. A. A. N., Abed, S. N., & Scholz, M. (2018). Wetlands for wastewater treatment and subsequent recycling of treated effluent: A review. *Environmental Science and Pollution Research International,* 25(24), 23595–23623. https://doi.org/10.1007/s11356-018-2629-3.

Arya, A. K., Joshi, K. K. J., Bachheti, A., & Deepti. (2020). A review on distribution and importance of wetlands in the perspective of India. *Journal of Applied and Natural Science,* 12(4), 710–720. https://doi.org/10.31018/jans.v12i4.2412.

Avila-Sanchez, J. S., Bates, C. E., Camacho, A. M., DiMaggio, A. M., Jacobson, S., Johnson, B. K., Page, M. T., Rhodes, E. C., Traub, N. J., & Perotto-Baldivieso, H. L. (2021). Rangeland ecology and management highlights. *Rangelands,* 43(3),127–130. https://doi.org/10.1016/j.rala.2021.06.001.

Bassi, N., Kumar, M. D., Sharma, A., & Pardha-Saradhi, P. (2014). Status of wetlands in India: A review of extent, ecosystem benefits, threats and management strategies. *Journal of Hydrology: Regional Studies,* 2, 1–19. https://doi.org/10.1016/j.ejrh.2014.07.001.

Boyd, C. E. (2020). Water quality protection. In *Water quality* (pp. 379–409). Springer. https://doi.org/10.1007/978-3-030-23335-8_18.

Boyd, J., & Banzhaf, S. (2007). What are ecosystem services? The need for standardized environmental accounting units. *Ecological Economics,* 63(2–3), 616–626. https://doi.org/10.1016/j.ecolecon.2007.01.002.

Brinson, M. M. (1993). A hydrogeomorphic classification for wetlands. U.S. Army Engineer Research and Development Center, Vicksburg, MS, USA. Technical Report WRP-DE-4. http://hdl.handle.net/11681/6483.

Broome, S. W., Craft, C. B., & Burchell, M. R. (2019). Tidal marsh creation. In *Coastal wetlands* (pp. 789–816). Elsevier. https://doi.org/10.1016/B978-0-444-63893-9.00022-8.

Byström, O., Andersson, H., & Gren, I. (2000). Economic criteria for using wetlands as nitrogen sinks under uncertainty. *Ecological Economics,* 35(1), 35–45. https://doi.org/10.1016/S0921-8009(00)00166-X.

Chandra, P., & Kumar, M. (2020). Contribution of microbes in the renovation of wetlands. In *Restoration of wetland ecosystem: A trajectory towards a sustainable environment* (pp. 101–124). Springer. https://doi.org/10.1007/978-981-13-7665-8_8.

Cowardin, L. M. (1979). Classification of wetlands and deepwater habitats of the United States. *Fish and Wildlife Service,* United States Department of the Interior. https://doi.org/10.5962/bhl.title.4108.

Cowardin, L. M., & Golet, F. C. (1995). US Fish and Wildlife Service 1979 wetland classification: A review. *Vegetation,* 118(1–2), 139–152. https://doi.org/10.1007/BF00045196.

Das, B. K., Parida, P. K., DebRoy, P., & Roy, A. (2022). *A situational analysis of small-scale inland open water fisheries in India: From vulnerability to viability.*

Davenport, M. A., Bridges, C. A., Mangun, J. C., Carver, A. D., Williard, K. W. J., & Jones, E. O. (2010). Building local community commitment to wetlands restoration: A case study of the Cache River wetlands in southern Illinois, USA. *Environmental Management,* 45(4), 711–722. https://doi.org/10.1007/s00267-010-9446-x.

Davidson, N. (2018). Wetland losses and the status of wetland-dependent species. In *The wetland book II: Distribution, description, and conservation* (pp. 369–381). Springer. https://doi.org/10.1007/978-94-007-4001-3_197.

Department of Fisheries, & Government of India. (2020). Pradhan Mantri Matsya Sampada Yojana. http://pmmsy.dof.gov.in/. Ministry of Fisheries, Animal Husbandry and Dairying p. 26.

Dworrak, T. V., Sauer, F. G., & Kiel, E. (2022). Wetland conservation and its effects on mosquito populations. *Wetlands,* 42(7), 1–14. https://doi.org/10.1007/s13157-022-01613-y.

Elliott, M., Burdon, D., Hemingway, K. L., & Apitz, S. E. (2007). Estuarine, coastal and marine ecosystem restoration: Confusing management and science–a revision of concepts. *Estuarine, Coastal and Shelf Science,* 74(3), 349–366. https://doi.org/10.1016/j.ecss.2007.05.034.

Ewart-Smith, J. L., Ollis, D. J., Day, J. A., & Malan, H. L. (2006). National wetland inventory: Development of a wetland classification system for South Africa. WRC Report no. KV 174/06." *Water Research Commission*. Pretoria.

Gebresllassie, Hagos, Gashaw, T., & Mehari, A. (2014). Wetland degradation in Ethiopia: Causes, consequences and remedies. *Environmental Earth Sciences,* 4(11), 40–48.

Ghermandi, A., van den Bergh, J. C. J. M., Brander, L. M., de Groot, H. L. F., & Nunes, P. A. L. D. (2008). The economic value of wetland conservation and creation: A meta-analysis. *SSRN Electronic Journal.* https://doi.org/10.2139/ssrn.1273002.

Giweta, M., & Worku, Y. (2018). Reversing the degradation of Ethiopian wetlands: Is it unachievable phrase or a call to effective action. *International Journal of Environmental Sciences and Natural Resources,* 14(5), 136–146.

Griffin, P. J., & Ali, S. H. (2014). Managing transboundary wetlands: The Ramsar Convention as a means of ecological diplomacy. *Journal of Environmental Studies and Sciences,* 4(3), 230–239. https://doi.org/10.1007/s13412-014-0173-0.

Groffman, P. M., & Crawford, M. K. (2003). Denitrification potential in urban riparian zones. *Journal of Environmental Quality,* 32(3), 1144–1149. https://doi.org/10.2134/jeq2003.1144.

Jain, C. K., Singhal, D. C., & Sharma, M. K. (2007). Estimating nutrient loadings using chemical mass balance approach. *Environmental Monitoring and Assessment,* 134(1–3), 385–396. https://doi.org/10.1007/s10661-007-9630-5.

Janse, J. H., Van Dam, A. A., Hes, E. M. A., de Klein, J. J. M., Finlayson, C. M., Janssen, A. B. G., van Wijk, D., Mooij, W. M., & Verhoeven, J. T. A. (2019). Towards a global model for wetlands ecosystem services. *Current Opinion in Environmental Sustainability,* 36, 11–19. https://doi.org/10.1016/j.cosust.2018.09.002.

Jarzemsky, R. D., Burchell, M. R., & Evans, R. O. (2013) The impact of manipulating surface topography on the hydrologic restoration of a forested coastal wetland. *Ecological Engineering,* 58, 35–43. https://doi.org/10.1016/j.ecoleng.2013.06.002.

Johnes, P. J., Gooddy, D. C., Heaton, T. H., Binley, A., Kennedy, M. P., Shand, P., & Prior, H. (2020). Determining the impact of riparian wetlands on nutrient cycling, storage and export in permeable agricultural catchments. *Water,* 12(1), 167. https://doi.org/10.3390/w12010167.

Kathiresan, K., & Thakur, S. (2008). Mangroves for the future: National strategy and action plan, India. *Ministry of Environment and Forests [Revised Draft].*

Kattel, G. R., Eyre, B. D., & Gell, P. A. (2020). Integration of palaeo-and-modern food webs reveal slow changes in a river floodplain wetland ecosystem. *Scientific Reports,* 10(1), 12955. https://doi.org/10.1038/s41598-020-69829-8.

Kingsford, R. T., Basset, A., & Jackson, L. (2016). Wetlands: Conservation's poor cousins. *Aquatic Conservation: Marine and Freshwater Ecosystems, 26*(5), 892–916. https://doi.org/10.1002/aqc.2709.

Kotze, D., Marneweck, G., Batchelor, A., Lindley, D., & Collins, N. (2009). A technique for rapidly assessing ecosystem services supplied by wetlands. *Mondi Wetland Project*.

Kumar, M. D., Bassi, N., Sivamohan, M. V. K., & Venkatachalam, L. (2013a). Agriculture in West Bengal: Can the new policies trigger a second Green Revolution? *Review of Development and Change,* 18(1), 37–51. https://doi.org/10.1177/0972266120130103.

Kumar, M. D., Panda, R., & Niranjan, V. (2012). Technology choices and institutions for improving the economic and livelihood benefits from multiple-use tanks in western Odisha. In *Water management, food security and sustainable agriculture in developing economies* (pp. 138-163). Routledge.

Kumar, M. D., & Vedantam, N. (2016). Groundwater use and decline in tank irrigation? Analysis from erstwhile Andhra Pradesh. In *Rural water systems for multiple uses and livelihood security* (pp. 145–182). Elsevier. https://doi.org/10.1016/B978-0-12-804132-1.00007-X.

Kumar, M. D., Vedantam, N., Bassi, N., Puri, S., & Sivamohan, M. V. K. (2012) "*Making rehabilitation' Work: protocols for improving performance of irrigation tanks in Andhra.*"

Kumar, M., Singh, N. K., & Singh, R. (2021). Application of constructed wetlands in degradation and detoxification of industrial effluents: Challenges and prospects. *Bioremediation for Environmental Sustainability,* 43–82. https://doi.org/10.1016/B978-0-12-820318-7.00003-4.

Lal, R. (2008). Carbon sequestration. Philosophical Transactions of the Royal Society B, 363(1492), 815–830. https://doi.org/10.1098/rstb.2007.2185.

Leykun, A. (2003). *The distribution and status of Ethiopian Wetlands: An overview, proceeding of a seminar on the resources and status of Ethiopia's wetlands.*

Mahdavi. Sahel, Bahram Salehi, jean granger, Meisam Amani, Brian Brisco, and Weimin Huang. "Remote sensing for wetland classification: A comprehensive review." *GIScience & Remote Sensing* 55, 5. (2018): 623–658. https://doi.org/10.1080/15481603.2017.1419602.

Mengesha, T. A. (2017). Review on the natural conditions and anthropogenic threats of Wetlands in Ethiopian. *Global Journal of Ecology,* 2(1), 6–14. https://doi.org/10.17352/gje.000004.

Mitsch, W. J., & Gosselink, J. G. (2007). *Wetlands.* John Wiley & Sons, Inc. Hoboken, NJ.

Morris, K., & Reich, P. (2013). *Understanding the relationship between livestock grazing and wetland condition Arthur Rylah Institute for Environmental Research Technical Report Series 252.*

Mu, C., Lu, Huicui, Wang, B., Bao, X., & Cui, W. (2013). Short-term effects of harvesting on carbon storage of boreal Larix gmelinii–Carex schmidtii forested wetlands in Daxing'anling, northeast China. *Forest Ecology and Management,* 293, 140–148. https://doi.org/10.1016/j.foreco.2012.12.031.

Nanson, G. C., & Croke, J. C. (1992). A genetic classification of floodplains. *Geomorphology,* 4(6), 459–486. https://doi.org/10.1016/0169-555X(92)90039-Q.

Neiland, A. E. (2008). Tropical river fisheries valuation: Background papers to a global synthesis. *WorldFish.*

Ollis, D. J., Ewart-Smith, J. L., Day, J. A., Job, N. M., Macfarlane, D. M., Snaddon, C. D., Sieben, E. J. J., Dini, J. A., & Mbona, N. (2015). The development of a classification system for inland aquatic ecosystems in South Africa. *Water SA*, 41(5), 727–745. https://doi.org/10.4314/wsa.v41i5.16.

Osland, M. J., Gabler, C. A., Grace, J. B., Day, R. H., McCoy, M. L., McLeod, J. L., From, A. S., Enwright, N. M., Feher, L. C., Stagg, C. L., & Hartley, S. B. (2018). Climate and plant controls on soil organic matter in coastal wetlands. *Global Change Biology,* 24(11), 5361–5379. https://doi.org/10.1111/gcb.14376.

Pant, Hari K., Rechcigl, J. E., & Adjei, M. B. (2003). Carbon sequestration in wetlands: Concept and estimation. *Journal of Food, Agriculture and Environment,* 1(2), 308–313.

Prasad, S. N., Ramachandra, T. V., Ahalya, N., SenGupta, T., Kumar, A., Tiwari, A. K., Vijayan, V. S., & Vijayan, L. (2002). Conservation of wetlands of India-a review. *Tropical Ecology,* 43(1), 173–186.

Pueyo-Ros, J., Ribas, A., & Fraguell, R. M. (2019). A cultural approach to wetlands restoration to assess its public acceptance. *Restoration Ecology,* 27(3), 626–637. https://doi.org/10.1111/rec.12896.

Qiu, J. (2019). Effects of landscape pattern on pollination, pest control, water quality, flood regulation, and cultural ecosystem services: A literature review and future research prospects. *Current Landscape Ecology Reports,* 4(4), 113–124. https://doi.org/10.1007/s40823-019-00045-5.

Rai, P. K. (2008). Heavy metal pollution in aquatic ecosystems and its phytoremediation using wetland plants: An ecosustainable approach. *International Journal of Phytoremediation,* 10(2), 131–158. https://doi.org/10.1080/15226510801913918.

Reid, A. J., Carlson, A. K., Creed, I. F., Eliason, E. J., Gell, P. A., Johnson, P. T. J., Kidd, K. A., MacCormack, T. J., Olden, J. D., Ormerod, S. J., Smol, J. P., Taylor, W. W., Tockner, K., Vermaire, J. C., Dudgeon, D., & Cooke, S. J. (2019). Emerging threats and persistent conservation challenges for freshwater biodiversity. *Biological Reviews of the Cambridge Philosophical Society,* 94(3), 849–873. https://doi.org/10.1111/brv.12480.

SAC. (2011). National wetland atlas. Space application center (SAC). *Indian Space Research Organization.*

Satyaveer, N. K. S., Rajanna, K.B., Rathore, S. S. & Verma, D.K. (2022). Status of Ramsar sites in *India The Pharma Innovation Journal.* 2022; 11(11S): 2088-2094.

Semeniuk, C. A., & Semeniuk, V. (1995). A geomorphic approach to global classification for inland wetlands. In. *Vegetatio.* Springer, 118(1–2), 103–124. https://doi.org/10.1007/BF00045193.

Sulaiman, A. A., Sulaeman, Y., & Minasny, B. (2019). A framework for the development of wetland for agricultural use in Indonesia. *Resources,* 8(1), 34. https://doi.org/10.3390/resources8010034.

Tan, L., Ge, Z., Zhou, X., Li, S., Li, X., & Tang, J. (2020). Conversion of coastal wetlands, riparian wetlands, and peatlands increases greenhouse gas emissions: A global meta-analysis. *Global Change Biology,* 26(3), 1638–1653. https://doi.org/10.1111/gcb.14933.

Verhoeven, J. T. A., Arheimer, B., Yin, C., & Hefting, M. M. (2006). Regional and global concerns over wetlands and water quality. *Trends in Ecology and Evolution*, 21(2), 96–103. https://doi.org/10.1016/j.tree.2005.11.015.

Wagner, K. I., Gallagher, S. K., Hayes, M., Lawrence, B. A., & Zedler, J. B. (2008). Wetland restoration in the new millennium: Do research efforts match opportunities? *Restoration Ecology,* 16(3), 367–372. https://doi.org/10.1111/j.1526-100X.2008.00433.x.

Wondie, A. (2012). Biodiversity and ecosystem services of lake TANA wetlands, Ethiopia. *Biodiversity Conservation and Ecosystem Services for Climate Change Mitigation and Sustainable Development,* 20, 91.

Xu, T., Weng, Baisha, Yan, D., Wang, K., Li, X., Bi, Wuxia, Li, M., Cheng, X., & Liu, Yinxue. (2019). Wetlands of international importance: Status, threats, and future protection. *International Journal of Environmental Research and Public Health,* 16(10), 1818. https://doi.org/10.3390/ijerph16101818.

Yitaferu, B., Hurni, H., & Zeleke, G. (2004). Processes of land degradation in the Lake Tana Basin (LTB), Amhara region, Ethiopia. In *Proceeding of the Symposium on Lake Tana watershed management* (pp. 5–12). Bahir Dar University.

Zhao, Q., Bai, J., Huang, Laibin, Gu, Binhe, Lu, Qiongqiong, & Gao, Z. (2016). A review of methodologies and success indicators for coastal wetland restoration. *Ecological Indicators,* 60, 442–452. https://doi.org/10.1016/j.ecolind.2015.07.003.

Chapter 9

Climate Smart Agriculture: A Roadmap to Sustainable Agriculture

Amrita Kumari
Priyanka Soni
Fahami Zaya
Rahul Kumar
Nidhi Singh
Riya Gupta
and Nandita Ghoshal*
Centre of Advanced Study in Botany, Department of Botany,
Institute of Science, Banaras Hindu University, Varanasi, India

Abstract

Agriculture systems globally are predicted to be altered drastically because of future climate change, resulting in not only posing challenges to food security, and greenhouse gas (GHG) emissions but also increasing uncertainty and vulnerability to various livelihoods. Agricultural systems have become very important not only for their role in achieving food security but also in mitigation of the climate change. Climate-smart agriculture (CSA) has been advocated as a strategy to address the challenges of climate change and food security by sustainably increasing productivity, strengthening resilience, and reducing GHG emissions, thereby achieving national food security and sustainable development goals. CSA has three major objectives viz.

* Corresponding Author's Email: n_ghoshal@yahoo.co.in.

In: Conserving, Halting and Reinstating Degraded Natural Resources …
Editors: Rajani Srivastava and Pankaj Kumar
ISBN: 979-8-89113-067-8

adaptation, mitigation, and productivity, which can be achieved by a combination of strategies that have traditionally been applied in the areas of conservation, climate change, environmental studies, ecology and agricultural farms along with financial assistance. Some of the important approaches to achieving the goals of CSA are effective resource management, advancing integrated renewable energy technology in agricultural systems, technical knowledge of farmers, promoting the role of institutions in CSA improvement, crop genetic modification investment, improving land use management, changing crop season, crop relocation, effective pest control, creating a forecasting system, and or using crop modeling techniques. The major challenge of CSA in the present scenario is to bridge the science-policy gap by creating an appropriate combination of strategies that can maximize synergies among the CSA components as well as avoid conflicting policies and negative trade-offs, especially with limited resources and competing food security priorities. Cross-disciplinary and multi-objective approaches in CSA are needed to strengthen the interface of the ecological, social, management, and economic dimensions thereby supporting maximum participation by various stakeholder groups around the world.

Keywords: climate change, climate smart agriculture, mitigation, resilience, sustainable agriculture

1. Introduction

The agricultural systems are facing the global challenge of increasing food production while reducing greenhouse gas emissions caused by food production. Climate change not only affects agricultural productivity but also contributes considerably to global greenhouse gas (GHG) emissions and is estimated to be about 11% of total anthropogenic GHG emissions not including land use change (Pinto et al., 2020; Smith et al., 2014). It has become more critical because of the given projections, that global poverty could increase by anywhere between 35 to 122 million people by 2050 under climate change scenarios, largely due to negative impacts on agricultural incomes. Agriculture systems around the world are predicted to be altered drastically as a result of future climate change, resulting in not only posing challenges to global food security but also increasing uncertainty and vulnerability to various livelihoods (Barasa et al., 2021; Morton, 2007). These problems will be aggravated shortly due to other related issues such as population increase, economic development, urbanization, and recurrent incidences of natural hazards such as variability in temperature and rainfall, drought, floods, etc. Agricultural systems have become very important not only for their role in achieving food security but also in mitigation of the climate change. Thus, sustainable management of agriculture has become an essential requirement of sustainable development (Paustian et al., 2016).

Modeling studies indicated that climate change would reduce and/or limit food production potential, globally and notably in many tropical and mid-latitudinal countries (Rosenzweig, 2014). Global demand for agricultural products including food, fiber, and fuel, continues to increase, mainly due to population growth, changes in dietary style resulting from an increase in per capita income and the need for alternative energy sources. It is estimated that global food production needs to increase considerably to feed 9.6 billion people by 2050. Moreover, land resource is finite so the availability of additional land for agricultural expansion is very limited. As a result, there will be intense pressure on available agricultural land that is already overburdened, especially in developing and underdeveloped countries, to produce more under changing climatic conditions and risks posed by other stresses.

Consequently, various international institutions came forward with the World Bank and the Food and Agricultural Organization (FAO) groups to design strategies for agricultural systems that can increase and improve food production from global to local levels. This resulted in the recognition of climate-smart agriculture as a global paradigm shift, not only by researchers but also by other stakeholders such as farmers, policymakers and investors, across private, public, and civil society sectors (Mora et al., 2015; Pugh et al., 2016).

Climate-smart agriculture has been promoted as a systematic approach to transform and protect the agriculture sector by developing agricultural strategies to ensure sustainable food security in the context of climate change (Strohmaier et. al., 2016). Climate-smart agriculture is described as a strategy to address the challenges of climate change and food security by sustainably increasing productivity, strengthening resilience, reducing GHG emissions, and increasing chances of achievement of national food security and sustainable development goals.

2. Historical Background

Climate-smart agriculture signifies a combination of strategies that have traditionally been applied in the areas of conservation, climate change, environmental studies, ecology and agricultural farms. However, the connection between these factors especially the agricultural systems and climate change is less studied and poorly understood. By 2007, IPCC and policy reviews of development agencies recognized the nature and extent of inter-relationship between the climate change and agriculture leading to the conclusion that 'there is global urgency that GHG emissions (CH_4 and N_2O) from agricultural lands were increasing and also there are interactions between mitigation and adaptation in the agricultural sector, which may occur simultaneously, but differ in their spatial and geographic characteristics.' It was obvious from the 2007–2008 global food crises that food security became an important issue for the masses, whereas resilience of agricultural production systems required a new perspective globally and especially in developing countries, to address the multiple interlinked challenges. In 2010, the FAO and World Bank officially recognize climate-smart agriculture as an approach to developing resilient agriculture systems which can provide solutions to climate change. The first global policy conference on climate smart agriculture was held at the Hague, Netherlands (31 October–5 November 2010), entitled, "Global Conference on Agriculture, Food Security and Climate Change" to formulate a roadmap for action organized by governments of developing and developed countries, along with not only the World Bank and FAO but also other stakeholders representing governments, international and regional organizations, the private sector, NGOs, philanthropic foundations, and the scientific communities (Neufeldt et al., 2013). For further development of the approach and partnership on climate-smart agriculture, conferences and workshops were held in the Netherlands (2011), Vietnam (2012), South Africa (2013), the USA (2013) and France (2015). In September 2014, during the UN Secretary-General's Climate Summit, the Global Alliance for climate-smart agriculture (GACSA) was launched with an emphasis on developing knowledge and coordinating global-level collaboration. "Policies, Practices and Financing for Food Security, Adaptation and Mitigation," was used to describe climate-smart agriculture in the original document of the FAO. Climate-smart agriculture thus was represented at the interface between science and policy-making and attempts to encourage action at the grassroots levels and also the mobilization of finances. The climate-smart agriculture approach promotes coordinated actions by farmers, researchers, the private sectors, civil societies, non-government organizations and policymakers towards climate-resilient pathways through four main action plans i.e., (a) building evidence; (b) increasing local institutional effectiveness; (c) fostering coherence between climate and agricultural policies; (d) linking climate and agricultural financing (Lipper et al., 2014).

3. Objectives of Climate Smart Agriculture

The phrase "climate adaptation and mitigation" was coined in 2009 to describe a method for adopting new agricultural methods to increase productivity, adaptation, and mitigation to face climate change (Lipper et al., 2014). According to the Food and Agriculture Organization of the United Nations, climate-smart agriculture refers to agricultural practices that increase productivity, contribute to increased resilience (adaptation), and reduce greenhouse-gas emissions (mitigation) where appropriate, while also assisting in the achievement of food security and sustainable development goals. Climate-smart agriculture has become known for the so-called "triple win," i.e., concurrently striving to achieve its three objectives or more popularly known as three pillars: "adaptation, mitigation, and food security (or productivity)" (Tasnoova & Iwamoto, 2009; Fanen & Olaleka, 2014). The three pillars of climate-smart agriculture are interconnected and represented in Figure 1

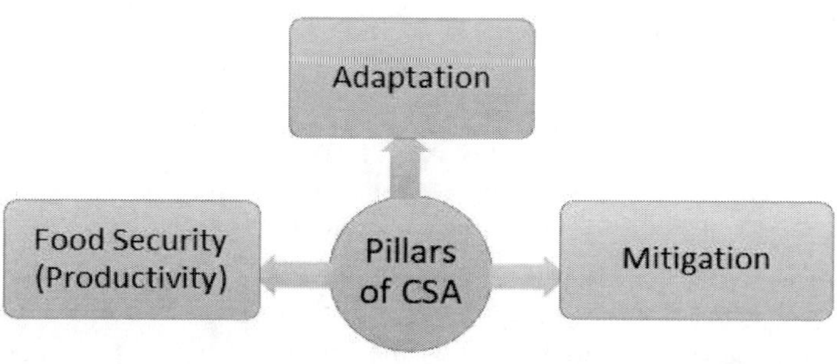

Figure 1. Representing three main pillars of climate-smart agriculture.

3.1. Productivity

Climate-smart agriculture aims to develop agricultural practices that increase productivity and income from crops, livestock, and fisheries while minimizing negative environmental impacts. It also contributes to the improvement of food and nutritional security (Branca et al., 2011). For maximum utilization of natural resources, sustainable intensification is a key component of climate-smart agriculture. Because of its vulnerability to the negative effects of climate change, crop farming is highly reliant on weather conditions. Increasing adaptive ability can aid in long-term intensification and, as a result, higher productivity (Yassin et al., 2008).

3.2. Adaptation

Climate-smart agriculture attempts to reduce farmers' exposure to short-term hazards while simultaneously building their resilience by improving their ability to adjust to long-term challenges. A special emphasis is placed on conserving the ecosystem that farmers and others rely on (Zougmore et al., 2016). Climate-smart strategies are required to improve climate change adaptation capabilities while also enhancing productivity. Climate-smart agriculture also ensures the system's adaptability to stresses and recovery (Basche, 2015). The integration of resources is focused on adaptation to develop activities that minimize negative effects and maximize positive benefits.

3.3. Mitigation

Climate-smart agriculture also seeks to reduce or eliminate Greenhouse gas emissions from food, fire, and fuel. It controls soil and vegetation in such a way that it can act as a carbon sink, absorbing CO_2 from the atmosphere. Mitigation, according to the FAO, is the ability of systems, societal groups, or individuals to protect, prevent, minimize, alleviate, or cope with risk while also recovering from pressures. For a system to be resilient in the face of stress over time, it must have high adaptive capacity.

These three pillars combine financial, social, and environmental concerns into a strategy that deals with food security and climate change issues simultaneously (Hafezi et al., 2017) by helping in adapting to climate change, increasing production, sustainability, resilience, and reducing greenhouse gas emissions (Mutamba & Mugoya, 2014).

Climate-smart agriculture is a method for achieving long-term agricultural improvement by combining scientific principles with financial assistance. Adaptation is a process that involves gaining control over hazards posed by global warming, such as variations in weather patterns (Alam et al., 2017). For scientists, program supervisors, and planners who are aware that climate change is real and detrimental to the community and ecological sustainability, variation in response to global warming is a critical concern.

In the twenty-first century, sustainability is an important idea for world ecological planning and decision-making (Braimoh & Osaki, 2010). Successful adaptation could address the most pressing challenges of global warming at a cost that is likely to be lower than the costs of development.

4. Challenges of Climate Smart Agriculture

Climate-smart agriculture is a good way to boost agricultural productivity and feed people all around the world in the face of changing climate conditions. It has arisen as a paradigm for capturing the idea that agricultural systems may be developed and implemented at the same time to improve food security and rural livelihoods, as well as to promote climate change adaptation and bring extra advantages. Despite the possibility of achieving a sustainable agricultural development goal by correctly implementing climate-smart agriculture, there is a poor linkage between climate-smart agriculture components at the field level. More critically, different tiers of stakeholders may not fully get an accurate perception. A deeper knowledge of the concept of climate-smart agriculture and its connections to sustainable agriculture as well as the path to promote climate-smart agriculture will allow for a more integrated climate change response and broader adaption of climate-smart agriculture methods to meet food security and sustainable development goals. As a result, it is critical to reinforce and promote climate-smart agriculture practices at the field level to achieve a more sustainable and prosperous future. However, for climate-smart agriculture to make a significant mark, major investments and coordination among various stakeholders are needed to implement the holistic concept of climate-smart agriculture widely (Pinto et al., 2020).

5. Linkages between Climate Smart Agriculture and Sustainable Agriculture

Environment, society and economy are three important components of sustainable agriculture that contribute to establishing an eco-friendly environment, economic profitability, and social equality (Newell & Taylor, 2018). Climate-smart agriculture aids the farmer in sustaining livelihoods and mitigating climate change to ensure food security by incorporating a resilience approach. Climate-smart agriculture is often represented as:

Climate-smart agriculture = Sustainable agriculture + Resilience – Emissions.

Agricultural sustainability considers current needs without compromising future generations' ability to meet those needs. Natural and human resources as well as social capacities such as rural society requirements, working and living conditions of workers, and people's well-being and safety, should be prioritized equally for the present and future (Managi & Karemera, 2005). By considering sustainability, climate-smart agriculture also focuses on animal welfare in agricultural enterprises, agroecosystems, and food systems. An agroecosystem is part of a system that connects the farmer's field to the consumer's hand. For a healthier agroecosystem, a food system approach necessitates education, research, and action. Agrobiodiversity, on the other hand, is concerned with the agroecosystem, which allows for the determination of the impact on the human community and its organizations on cultivation and ecological sustainability (Braimoh & Oskai, 2010). Climate-smart agriculture is focused not only on practices and related technology but also on the consequences at the farm level for encouraging initiatives that address the risks of productivity, adaptability, and mitigation all at the same time (Brodt et al., 2011). As a result, climate-smart agriculture frequently emphasizes the interconnections between production, adaptation, and mitigation as well as the relationships that emerge at many levels, such as broader economic and socio-ecological ramifications (Precious et al., 2017). As a result, multiple stakeholders must devote sufficient resources to transform conventional agriculture into sustainable agriculture via climate-smart agriculture methods (Mwongera et al., 2017).

5.1. Major Approaches of Climate Smart Agriculture for Sustainable Agriculture

5.1.1. Effective Resource Management

Climate-smart agriculture and the future environment rely heavily on resource management. It is estimated that over one-third of all food is lost owing to inefficient resource management (Hartter et al., 2017). All aspects of the food supply chain, including cultivation, transport, preservation, handling, food preparation, and consumption, are viable candidates for improving resource management efficiency. Approximately 90% of the wood that has been removed is utilized to make domestic items as well as food preparation. Deforestation can be reduced by efficient resource management in food preparation, such as using a cook stove.

5.1.2. Advancing Integrated Renewable Energy Technology in Agricultural Systems

To reduce GHG emissions, appropriate technological innovation in agriculture is required. Wind generators, home solar panels, photovoltaic lights, biogas exclusion units, resources for bio-oil exploration and filtering, information and communication technology innovation, and other associated practices are all vital new technological advancements for the environment (Bochtis et al., 2014; Basche, 2015). This current technical innovation has the potential to add value to the industrial process.

5.1.3. Technical Knowledge of Farmers

Most farmers, particularly in Asia and Africa are poor and have limited resources but they have been experimenting with environmental variations for hundreds of years. They have a wide range of experience with a variety of approaches to dealing with environmental issues. This level of knowledge is essential to meet today's demands but they continue to struggle with correctly applying and adopting climate-smart agriculture, despite increased concern at national and international policy levels about the sustainability of agricultural development and food security in developing countries. Lipper et al., (2014) found that farmers in rural areas are unaware of and uninformed about the usage of these strategies. The adoption of climate-smart agriculture can be viewed as a process in which farmers fully embrace environment-friendly agriculture techniques as their best and only option. The adoption process consists of five steps: awareness, interest, assessment, trial, and adoption. In practice, this may not always be the case because interest or willingness may come earlier. The level of climate-smart agriculture adoption may vary among people, societies and households.

5.1.4. Promoting the Role of Institutions in Climate-Smart Agriculture Improvement

Adoption of the climate-smart agriculture technique requires strong institutional backing. Institutions play an important influence in both the transfer of new technologies and technological innovation. Farmers' field educational institutes, for example, can assist in the adoption of climate-smart technologies (Alam et al., 2018b). Because climate-smart agriculture aims to improve people's and ecosystems' livelihoods through a better combination of technologies and practices that take natural hazards into account on a local and international scale, it requires institutional support such as information, innovation, investment, and insurance to reap the full benefits of this approach.

5.1.5. Crop Genetic Modification Investment

Crop genetic modification is a climate-smart agriculture strategy that helps to boost crop variety productivity. Every 25 years, a breakthrough in crop research occurs, resulting in nearly doubled productivity for feeding the world's growing population (Basche, 2015). By providing novel plant kinds that are resistant to heat and salt stress, genetic modification procedures can be improved while reducing the risks of environmental conditions. Increased crop yields, lower costs for food or drug production, less need for pesticides, improved nutrient composition and food quality, pest and disease resistance, greater food security, and medical benefits for the world, and expanding population are a few advantages of genetic engineering in agriculture (Phillips, 2008). Currently, the pipeline for GM crops has been extended to include additional fruits, vegetables, and cereals like lettuce, strawberries, eggplant, sugarcane, rice, wheat, carrots, etc. These crops are expected to be used to increase vaccine production, provide nutrients for animal feed, and confer salinity and drought resistance traits to help plants grow in unfavorable environments and climates. Since they were first made available for commercial use, GM crops have benefited the economy and the environment (Bawa et al., 2013). Over a relatively small region, the world food crop yield increased by 370 million tonnes between 1996 and 2013 (Zhang et al., 2016).

5.1.6. Improving Land Use Management

Proper land use planning such as determining the location of plant and animal production and adjusting the concentration of plant feeds and bug sprays, can assist to mitigate the effects of global warming on agriculture (Ahmad et al., 2014; Basche, 2015). On the earth, mankind has used around 26% (3.3 billion ha) of the land for pasture and agriculture (Tong et al., 2003). Due to an increase in the already high population density in areas with fertile soil, there has been an increase in land pressure in Asia over the past 40 years (Bringezu et al., 2012). The results of extensive research on land use have increased recognition of the critical need to address land use management within studies of global environmental change (Deng et al., 2016; 2017). Crop yield may also be increased by using agricultural inputs like electricity, fertilizer, and machinery. To increase land eco-efficiency and advance sustainable development, the land use structure and its spatial dimensions should be planned logically. This will not only lead to improved land use but will also help to protect the environment for a sustainable society (Deng and Gibson, 2020).

5.1.7. Changing Crop Season

Cultivating high-yield varieties, increasing agricultural intensity, and adapting to diverse hydrological circumstances are all examples of modified agriculture systems. The growing season can be altered to prevent sterility caused by high temperatures allowing the flowering interval to be synchronized with the most recent time (Arslan et al., 2015). Shifting the planting or planting time to acquire the interval and avoid harsh weather events throughout the year can assist to mitigate the climate change negative effects that are common in semi-arid tropics and dry places (Basche, 2015).

5.1.8. Crop Relocation

Crop relocation refers to separating plants and varieties according to their climate sensitivity. The climate-smart agriculture technique allows crops to be relocated according to their climatic sensitivity during their vegetative and reproductive stages. It requires a particular quantity of daylight, temperature, and humidity to grow vegetatively and reproduce. As a result, the sowing and transplanting periods of crops should be favorable to support consistent plant growth. Similarly, harvesting time should be considered to ensure proper harvesting and reduce harvesting loss (Baba et al., 2017). Furthermore, some plants, such as tea, apple, and pear, are closely associated with prolonged periods of heavy rain or hot, dry weather.

5.1.9. Effective Pest Control

The climate-smart agriculture allows pest management techniques such as integrated pest management (IPM) to be used in an environmentally sustainable manner. IPM is an integrated approach to pest control that includes physical methods such as light traps, nets, and polyethylene, as well as biological methods such as the use of various botanical bio-controls and bio-agents. Climate-smart agriculture encourages the development of pest-resistant cultivars.

5.1.10. Creating a Forecasting System

Forecasting is an early warning system for climate change, pest attacks, harvesting time, and yield. Weather forecasting and early warning systems help reduce the danger of environmental disasters. This way farmers will be prepared for the worst situation and would be able to protect the crops from damage.

5.1.11. Using Crop Modeling Techniques

Crop modeling is a novel and recently developed tool for decreasing agricultural hazards. Models can help identify potential dangers from changing environmental conditions to future crop development (Neufeldt et al., 2013). Crop simulation designs are crucial in field tests for developing new crop control approaches. Simulation modeling and integrated assessment modeling are two such techniques.

a. *Simulation modeling:* Simulation modeling is used to examine the effects of different climate-smart agriculture interventions on crop, livestock, and farming system level. In India, it has been used to assess the effectiveness of agronomic practices in northeastern India (Shirsath et al., 2017).
b. *Integrated assessment modeling:* For prioritization, a variety of integrated evaluation approaches have been used. The International Model for Policy Analysis of Agricultural Commodities and Trade (IMPACT) model is one of them which estimates welfare indicators like the availability of nutrients, the population at risk of hunger, and the number of undernourished children in developing countries (Robinson et al., 2015).

5.1.12. Modeling and Forecasting Integration

Plant simulator designs, if correctly updated and confirmed in field tests, can efficiently compute and analyze the effects of specific standard water stress conditions on plant efficiency. The plant designs allow for adjustments in ecological parameters such as temperature and imitate plant responses. Due to the complexity of the issue, research and development are required to forecast crop requirements.

5.1.13. Use of Organic Fertilizers

The use of organic fertilizers like biochar can be very useful for the soil as well as the environment. It would help in more carbon sequestration and thus can be a very promising example of climate-smart agriculture. Some reports also showed the effects of various organic fertilizers such as plant-based, animal-based, rock-based fertilizers and liquid organic fertilizers on the growth and yield of crops and vegetables to ensure climate-smart agriculture (Hameed et al., 2019; Thanaporn & Nuntavun, 2019).

Conclusion

Climate-smart agriculture is beneficial not only for agricultural economic returns and long-term sustainable food production but also for addressing the vulnerabilities of climate change in a better way. Agriculture production systems can be made more resilient and resource-efficient by promoting climate-resilient action approaches such as (1) building evidence; (2) increasing local institutional effectiveness; (3) fostering coherence between climate and agricultural policies; and (4) linking climate and agricultural financing. To achieve the full benefit of climate-smart agriculture, approaches must coordinate and integrate the efforts of farmers, researchers, corporates, civil societies and policymakers. However, most of the climate-smart agriculture progressions are still at the policy level. Further, the framework description needs to understand the current developments and activities conducted within the preview of the climate-smart agriculture research arena including synergies, trade-offs, and co-benefits of livelihood and environmental outcomes. Strategies of climate-smart agriculture must include optimal combinations of adaptation and mitigation strategies to ensure maximum socio-economic development co-benefits for the farmers, especially the smallholder farmers of underdeveloped or developing countries. The major challenge of climate-smart agriculture in the present scenario is to bridge the science-policy gap by creating an appropriate combination of strategies that can maximize synergies among the climate-smart agriculture components on one hand, and on the other, avoiding conflicting policies and negative trade-offs, especially with limited resources and competing food security priorities. Future research in climate-smart agriculture is needed to include cross-disciplinary and multi-objective approaches by strengthening the interface of the social, management, and economic dimensions of climate-smart agriculture and also by supporting maximum participation by various stakeholders across the globe.

Acknowledgments

We thank the Head, and the Coordinator, Centre of Advanced Study in Botany, Department of Botany, for providing the required facilities. The Faculty Incentive Grant under the Institute of Eminence Scheme provided to Prof. N. Ghoshal is gratefully acknowledged. The authors are grateful to University Grants Commission, New Delhi, India for University Research Fellowship to Ms. Priyanka Soni, Ms. Fahami Zaya, Ms. Amrita Kumari; Junior Research Fellowship to Ms. Nidhi Singh; Council for Scientific and Industrial Research, New Delhi, India, for Senior Research Fellowship to Mr. Rahul Kumar and Junior Research Fellowship to Ms. Riya Gupta.

References

Ahmad, M., Rajapaksha, A. U., Lim, J. E., Zhang, M., Bolan, N., Mohan, D., Vithanage, M., Lee, S. S., & Ok, Y. S. (2014). Biochar as a sorbent for contaminant management in soil and water: a review. *Chemosphere*, *99*, 19-33. https://doi.org/10.1016/j.chemosphere.2013.10.071.

Alam, G. M., Alam, K., & Mushtaq, S. (2017). Climate change perceptions and local adaptation strategies of hazard-prone rural households in Bangladesh. *Climate risk management*, *17*, 52-63. https://doi.org/10.1016/j.chemosphere.2013.10.071.

Alam, G. M., Alam, K., Mushtaq, S., Khatun, M. N., & Mamun, M. A. K. (2019). Influence of socio-demographic factors on mobile phone adoption in rural Bangladesh: Policy implications. *Information Development*, *35*(5), 739-748. https://doi.org/10.1177/0266666918792040.

Arslan, A., McCarthy, N., Lipper, L., Asfaw, S., Cattaneo, A., & Kokwe, M. (2015). Climate smart agriculture? Assessing the adaptation implications in Zambia. *Journal of Agricultural Economics*, *66*(3), 753-780. https://doi.org/10.1111/1477-9552.12107.

Barasa, P. M., Botai, C. M., Botai, J. O., & Mabhaudhi, T. (2021). A review of climate-smart agriculture research and applications in Africa. *Agronomy*, *11*(6), 1255-1282. https://doi.org/10.3390/agronomy11061255.

Basche, A. D. (2015). *Climate-smart agriculture in Midwest cropping systems: evaluating the benefits and tradeoffs of cover crops* (Doctoral dissertation, Iowa State University). https://search.proquest.com/openview/ea6c0e6bd6f3b48b46b26f68edba4ec5/1?pq-origsite=gscholar&cbl=18750.

Bawa, A. S., & Anilakumar, K. R. (2013). Genetically modified foods: safety, risks and public concerns - a review. *Journal of food science and technology, 50*(6), 1035-1046. https://doi.org/10.1007/s13197-012-0899-1.

Bochtis, D. D., Sørensen, C. G. C., & Busato, P. (2014). 'Advances in agricultural machinery management: a review.' *Biosystems Engineering*, *126*(10), 69–81. https://doi.org/10.1016/j.biosystemseng.2014.07.012.

Braimoh, A. K., & Osaki, M. (2010). 'Land-use change and environmental sustainability.' *Sustainability Science*, *5*(1), 5–7. https://doi.org/10.1007/s11625-009-0092-2, https://link.springer.com/article/10.1007/s11625-009-0092-2.

Branca, G., McCarthy, N., Lipper, L., & Jolejole, M. C. (2011). Climate-smart agriculture: a synthesis of empirical evidence of food security and mitigation benefits from improved cropland management. *Mitigation of climate change in agricultureseries*, *3*, 1-42. https://www.academia.edu/download/49470332/Climate_Smart_Agriculture_A_Synthesis_of20161009-11118-13ee6jm.pdf.

Bringezu, S., O'Brien, M., & Schütz, H. (2012). Beyond biofuels: Assessing global land use for domestic consumption of biomass: A conceptual and empirical contribution to sustainable management of global resources. *Land use policy*, *29*(1), 224-232. https://doi.org/10.1016/j.landusepol.2011.06.010.

Brodt, S., Six, J., Feenstra, G., Ingels, C., & Campbell, D. (2011). Sustainable agriculture. *National Education Knowledge*, *3*(1). https://charlesscd.com/wp-content/uploads/2019/03/agriculture-and-the-environment-2019-Pennsylvania-Envirothon.pdf.

Change, C. (2007). IPCC fourth assessment report. *The physical science basis, 2*, 580-595. http://sciencepolicy.colorado.edu/admin/publication_files/resource-2523-Alvord_ipcc.pdf.

De Pinto, A., Cenacchi, N., Kwon, H. Y., Koo, J., & Dunston, S. (2020). Climate smart agriculture and global food-crop production. *PLoS One*, *15*(4), e0231764. https://doi.org/10.1371/journal.pone.0231764.

Deng, X., & Gibson, J. (2020). Sustainable land use management for improving land eco-efficiency: A case study of Hebei, China. *Annals of Operations Research*, *290*(1), 265-277. https://doi.org/10.1007/s10479-018-2874-3.

Deng, X., Gibson, J., & Wang, P. (2017). Relationship between landscape diversity and crop production: a case study in the Hebei Province of China based on multi-source data integration. *Journal of Cleaner Production*, *142*, 985-992. https://doi.org/10.1016/j.jclepro.2016.03.174.

Hafezi, R., Bahrami, M., & Akhavan, A. N. (2017). Sustainability in development: rethinking about old paradigms. *World Review of Science, Technology and*

Sustainable Development, *13*(2), 192-204. https://doi.org/10.1504/WRSTSD.2017.084185.

Hammed, T. B., Oloruntoba, E. O., & Ana, G. R. E. E. (2019). Enhancing growth and yield of crops with nutrient-enriched organic fertilizer at wet and dry seasons in ensuring climate-smart agriculture. *International Journal of Recycling of Organic Waste in Agriculture*, *8*(1), 81-92. https://doi.org/10.1007/s40093-019-0274-6.

Hartter, J., Hamilton, L. C., Boag, A. E., Stevens, F. R., Ducey, M. J., Christoffersen, N. D., Oester, P. T., & Palace, M. W. (2018). Does it matter if people think climate change is human caused? *Climate Services*, *10*, 53-62. https://doi.org/10.1016/j.cliser.2017.06.014.

Kpadonou, R. A. B., Owiyo, T., Barbier, B., Denton, F., Rutabingwa, F., & Kiema, A. (2017). Advancing climate-smart-agriculture in developing drylands: Joint analysis of the adoption of multiple on-farm soil and water conservation technologies in West African Sahel. *Land use policy*, *61*, 196-207. https://doi.org/10.1016/j.landusepol.2016.10.050.

Lipper, L., Thornton, P., Campbell, B. M., Baedeker, T., Braimoh, A., Bwalya, M., Caron, P., Cattaneo, A., Garrity, D., Henry, K. & Torquebiau, E. F. (2014). Climate-smart agriculture for food security. *Nature climate change*, *4*(12), 1068-1072. https://www.nature.com/articles/nclimate2437.

Managi, S., & Karemera, D. (2005). The effects of environment and technology on agricultural export. *International journal of agricultural resources, governance and ecology*, *4*(1), 45-63. https://doi.org/10.1504/IJARGE.2005.006438.

Mora, C., Caldwell, I. R., Caldwell, J. M., Fisher, M. R., Genco, B. M., & Running, S. W. (2015). Suitable days for plant growth disappear under projected climate change: Potential human and biotic vulnerability. *PLoS biology*, *13*(6), e1002167. https://doi.org/10.1371/journal.pbio.1002167.

Morton, J. F. (2007). The impact of climate change on smallholder and subsistence agriculture. *Proceedings of the national academy of sciences*, *104*(50), 19680-19685. https://doi.org/10.1073/pnas.0701855104.

Mutamba, M., & Mugoya, M. (2014). *Climate-Smart Agriculture: Farmers' Perspectives*. 1-14. https://hdl.handle.net/10568/42433.

Mwongera, C., Shikuku, K. M., Twyman, J., Läderach, P., Ampaire, E., Van Asten, P., Steve Twomlow & Winowiecki, L. A. (2017). Climate smart agriculture rapid appraisal (CSA-RA): A tool for prioritizing context-specific climate smart agriculture technologies. *Agricultural systems*, *151*, 192-203. https://doi.org/10.1016/j.agsy.2016.05.009.

Neufeldt, H., Jahn, M., Campbell, B. M., Beddington, J. R., Declerck, F., De Pinto, A., & Zougmoré, R. (2013). Beyond climate-smart agriculture: toward safe operating spaces for global food systems. *Agriculture & Food Security*, *2*, 1–6. http://www.agricultureandfoodsecurity.com/content/2/1/12.

Newell, P., & Taylor, O. (2018). Contested landscapes: the global political economy of climate-smart agriculture. *The Journal of Peasant Studies*, *45*(1), 108-129. https://doi.org/10.1080/03066150.2017.1324426.

Paustian, K., Lehmann, J., Ogle, S., Reay, D., Robertson, G. P., & Smith, P. (2016). Climate-smart soils. *Nature*, *532*(7597), 49-57. https://www.nature.com/articles/nature17174.

Phibunwatthanawong, T., & Riddech, N. (2019). Liquid organic fertilizer production for growing vegetables under hydroponic condition. *International Journal of Recycling of Organic Waste in Agriculture*, *8*(4), 369-380. https://doi.org/10.1007/s40093-019-0257-7.

Phillips, T. (2008). Genetically modified organisms (GMOs): Transgenic crops and recombinant DNA technology. *Nature Education*, *1*(1), 213. https://7arrahmaan7od.wordpress.com/2014/05/25/gmos-genitaclly-modified-organisms-what-is-gmos/.

Pugh, T. A. M., Müller, C., Elliott, J., Deryng, D., Folberth, C., Olin, S., Schmidt, E., & Arneth, A. (2016). Climate analogues suggest limited potential for intensification of production on current croplands under climate change. *Nature communications*, *7*(1), 1-8. https://www.nature.com/articles/ncomms12608.

Robinson, S., Mason-D'Croz, D., Sulser, T., Islam, S., Robertson, R., Zhu, T., Gueneau, A., Pitois, G., & Rosegrant, M. W. (2015). *The international model for policy analysis of agricultural commodities and trade (IMPACT): model description for version 3*. http://dx.doi.org/10.2139/ssrn.2741234.

Rosenzweig, C., Elliott, J., Deryng, D., Ruane, A. C., Müller, C., Arneth, A., Kenneth J., Christian F., Michael G., Nikolay K., Kathleen N., Franziska P., Thomas A., Pugh, T. A. M., Schmid, E., Stehfest, E., Yang, H., & Jones, J. W. (2014). Assessing agricultural risks of climate change in the 21st century in a global gridded crop model intercomparison. *Proceedings of the national academy of sciences*, *111*(9), 3268-3273. https://doi.org/10.1073/pnas.1222463110.

Senyolo, M. P., Long, T. B., Blok, V., & Omta, O. (2018). How the characteristics of innovations impact their adoption: An exploration of climate-smart agricultural innovations in South Africa. *Journal of Cleaner Production*, *172*, 3825-3840. https://doi.org/10.1016/j.jclepro.2017.06.019.

Shirsath, P. B., Aggarwal, P. K., Thornton, P. K., & Dunnett, A. (2017). Prioritizing climate-smart agricultural land use options at a regional scale. *Agricultural Systems*, *151*, 174-183. https://doi.org/10.1016/j.agsy.2016.09.018.

Smith, P., Bustamante, M., Ahammad, H., Clark, H., Dong, H., Elsiddig, E. A., Haberl, H., Harper, R., House, J., Jafari, M., Masera, O., Mbow, C., Ravindranath, N., Rice, C. W., Abad, C. R., Romanovskaya, A., Sperling, F., Tubiello, F. & Bolwig, S. (2014). Agriculture, forestry and other land use (AFOLU). In *Climate change 2014: mitigation of climate change. Contribution of Working Group III to the Fifth Assessment Report of the Intergovernmental Panel on Climate Change* (pp. 811-922). Cambridge University Press. https://rucforsk.ruc.dk/ws/portalfiles/portal/81002708/ipcc_wg3_ar5_chapter11.pdf.

Strohmaier, R., Rioux, J., Seggel, A., Meybeck, A., Bernoux, M., Salvatore, JavierM.& Agostini, A. (2016). The agriculture sectors in the Intended Nationally Determined Contributions: analysis. *Environment and Natural Resources Management Working Paper*, *62*.

Tasnoova, S., & Iwamoto, I. (2009). The improvement of livelihood and rural development by the exotic Pangasiid catfish farming in Bangladesh. *World Review of Science, Technology & Sustainable Development*, *6*(1), 64-74. https://doi.org/10.1504/WRSTSD.2009.022458.

Terdoo, F., & Adekola, O. (2014). Assessing the role of climate-smart agriculture in combating climate change, desertification and improving rural livelihood in Northern Nigeria. *African journal of agricultural research*, *9*(15), 1180-1191. http://ray.yorksj.ac.uk/id/eprint/4109/.

Tong, C., Hall, C. A., & Wang, H. (2003). Land use change in rice, wheat and maize production in China (1961–1998). *Agriculture, Ecosystems & Environment*, *95*(2-3), 523-536. https://doi.org/10.1016/S0167-8809(02)00182-2.

Yassin, M. M., Tubail, K. M., & Al-Dadah, J. Y. (2008). Towards strategies for pollution control on the use of wastewater effluent in sustainable agriculture in the Gaza Strip. *World Review of Science, Technology & Sustainable Development*, *5*(1), 66-78. https://doi.org/10.1504/WRSTSD.2008.017813.

Zhang, C., Wohlhueter, R., & Zhang, H. (2016). Genetically modified foods: A critical review of their promise and problems. *Food Science & Human Wellness, 5*(3), 116-123. https://doi.org/10.1016/j.fshw.2016.04.002.

Zougmoré, R., Partey, S., Ouédraogo, M., Omitoyin, B., Thomas, T., Ayantunde, A., Ericksen, P., Said, M., & Jalloh, A. (2016). Toward climate-smart agriculture in West Africa: a review of climate change impacts, adaptation strategies and policy developments for the livestock, fishery and crop production sectors. *Agriculture & Food Security*, *5*(1), 1-16. https://agricultureandfoodsecurity.biomedcentral.com/articles/10.1186/s40066-016-0075-3.

Chapter 10

Grassland Degradation and Its Restoration

Mahesh Chalimalla
and Rajani Srivastava*
Department of Environment and Sustainable Development, IESD,
Rajiv Gandhi South Campus, Banaras Hindu University, Mirzapur, India

Abstract

Ecosystem services are irreplaceable natural resources whose value is directly correlated with both ecological change and human well-being. Nearly 20% of the world's land surface is covered by grasslands, one of the most widespread plant groups. Because of climate change and overgrazing, their topsoils are vulnerable to erosion, and Soil Organic Carbon (SOC) supplies are declining where there is severe erosion. Grasslands are crucial for raising cattle production and protecting the environment. The conversion and degradation of the grassland biome are mostly caused by agricultural intensification, plantation forestry, urban expansion, mining, invasive non-native flora, and rural sprawl. Monitoring grassland degradation may be more accurate and impartial than using existing environmental indexes. Through excessive grazing, improper use of crops, and unrestricted collection of fuel and medicinal plants, grassland degradation is spreading over large areas. Grasslands restoration techniques vary as per intended major use. We can restore grasslands by using ecological approaches like protection from grazing, soil and water conservation, controlled fire, bush clearing, application of fertilizers or by using grasses and legumes. For the restoration of grassland, we can use ecological, biological, or biotechnological tools. But the best option is integrated techniques of grassland management.

* Corresponding Author's Email: rajani.srivastava25@gmail.com.

In: Conserving, Halting and Reinstating Degraded Natural Resources …
Editors: Rajani Srivastava and Pankaj Kumar
ISBN: 979-8-89113-067-8

Keywords: grassland ecosystem, grassland degradation, climate factor, land use

1. Introduction

Grassland provides a wide range of ecosystem services and is one of the ecosystems with the highest species richness in the world. Grasslands as terrestrial ecosystems dominated by herbaceous and shrub vegetation (belonging to the family Poaceae, Cyperaceae and Juncaceae) and maintained by fire, drought and/or freezing temperature. Grasslands occur naturally all over the world and cover approximately 41 percent of Earth's land area, many grasslands are better characterized as "fragments" than as "vast expanses" (White et al., 2000). It is dominated by both perennial species (such as in northwest Europe) and annual species (in warmer climatic condition). The length of grasses varies from very short (less than 30 cm) to quite tall, as in the case of North American tallgrass prairie, south American grasslands and African savanna. It covers 69% of the world's agricultural land area and occupies around 40% of the planet's surface.

Grasslands play an important role in livestock farming and environmental conservation (Zhang et al., 2022). Grasslands provide hundreds of millions of people with cultural values, food, fuel, fiber, and medications. Grassland provides essential resources, such as food, meat, milk, skin, wool, and other livestock products for the economic expansion of animal husbandry (Lauenroth & Adler, 2008). Grasslands are crucial to the global carbon cycle because 90% of their biomass is underground, they accumulate carbon quickly, and they decompose organic matter slowly (Abdalla et al., 2018). They make it easier for water to seep or infiltrate into the soil and support the preservation of hydrological cycles. Finally, grasses enhance the beauty of many regions' landscapes. It takes almost 7 years at least to modify fertility in soils. In general, we can divide grasslands into six major types, these are- tropical and subtropical grasslands, temperate grasslands, flooded grasslands, montane grasslands, polar grasslands and xeric grasslands. From all types of grassland, we derive many benefits and ecological services.

2. Benefits and Ecological Services

Grasslands provide crucial ecological tasks including wind prevention and sand fixation, water purification, and biodiversity preservation in addition to providing key economic activities like animal products. Ecosystem services are described as "various outputs, conditions, and processes of natural biological systems which, in one way or another, directly or indirectly, benefit humans and significantly enhance their social welfare" (Zhang et al., 2016; Yan et al., 2011). These include a range of mechanisms by which grasslands generate a wealth of advantageous resources, such as food for ruminants, clean water by acting as an excellent watershed, preserving biodiversity by providing wildlife with suitable habitat, etc. Many Types of ecosystem services provided by grassland are given in Figure 1.

2.1. Biological Diversity and Habitat

Pollination is made easier in grasslands by their biological diversity. Because grassland flora and fauna have coevolved, many plants have developed specialized floral traits (Breidenbach et al., 2022).

2.2. Forage

Grasslands provide herbage for livestock in the form of grazing land. There have been many replacements for these traditional grasslands with less diversified grasses including both indigenous and exotic. The most recent grasslands have also been established to provide a habitat for wildlife and create corridors for animal movement in pastures (Bruinenberg et al., 2002). Provides spaces for nesting and brood-rearing, winter cover, and recreation areas for hunting and other outdoor pursuits (Fondell & Ball, 2004).

2.3. Recreation

For the general public, grasslands offer several recreational activities. Generally, Humans spend lots of time outdoors like bird watching and hunting. The fastest-growing ecotourism activity may be bird watching, which is also a significant source of income for the local public and communities (Bengtsson et al., 2019).

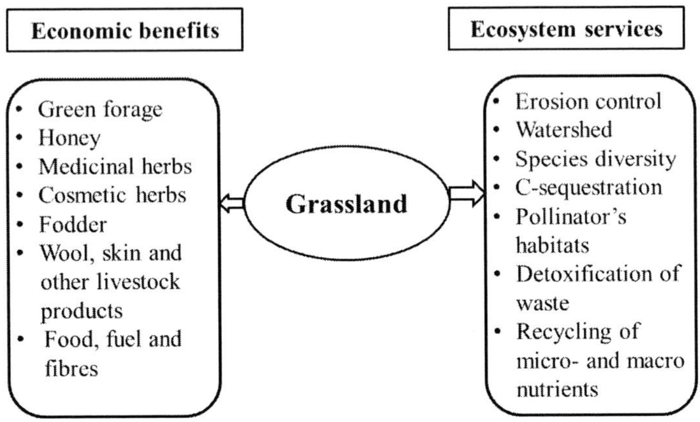

Figure 1. Multiple benefits and ecosystem services derived from grassland.

2.4. Other Benefits

In addition to the ecosystem services provided by forested areas, grasslands often provide a variety of other advantages, including the dispersal of seeds, the preservation of abundant and threatened plant species, drought and flood mitigation through effective microclimate maintenance, the recycling of macro- and micronutrients in the soil as the plant life cycle progresses, and the detoxification of various wastes through decomposition (Winfree et al., 2015; When et al., 2018).

3. Grassland Degradation

The degradation of the grassland ecosystem is caused by a variety of factors, including both natural and human-made ones. These include overgrazing, mowing, reclamation, cutting, and mineral mining, as well as long-term droughts, wind, water, temperature changes, and precipitation. Due to disruption from human activity and changes in the natural environment, global warming has increased by 0.2 °C/10 year during the past 30 years. Currently, Asia has the highest number of regions exhibiting fast land-cover change, especially grassland degradation. Several factors are responsible for the degradation of the grassland ecosystem like large bovine population, free grazing practices, lack of management practices, natural constraints like extreme temperature, variable precipitation, scarcity of moisture in arid and semi-arid situations, fire, urbanization, fragmentation etc.

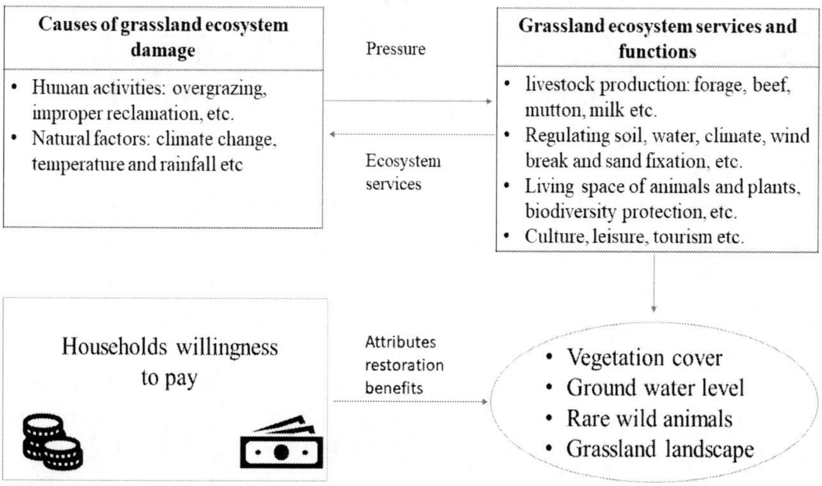

Figure 2. Causes of grassland ecosystem damage and services hampered provided by grasslands.

Degradation has been asserted by many researchers such as overgrazing by animals, improper agricultural management, collection of wood for fuel and herbs for medicine, and destruction by rodents (Figure 2). Although some aspects have improved, degradation continues to expand and intensify. Remote sensing is currently being used to monitor the degradation of grasslands, with most studies being focused on regional issues and less on national ones (Ali et al., 2016). Changes in the grassland seed bank and soil parameters, such as soil water content, soil organic carbon, total nitrogen, soil bulk density, soil microbes, soil enzyme, etc., can result in soil degradation and even.

Degradation of grasslands has the potential to disrupt the network structure of soil fungi, with negative repercussions for ecosystem multi-functionality and fungal community compositional stability (Andrade et al., 2015). The degradation of grasslands has a significant negative impact on the environment since they are essential for maintaining biodiversity and regulating the climate and biogeochemical cycles on a global scale.

Significant soil carbon loss can result from invasive species (Bardgett et al., 2021). Degradation endangers the services and functioning of local ecosystems (Gang et al., 2014). The demand for food has increased quickly due to population expansion, and grassland has been extensively converted to agriculture (Jiang et al., 2006). Vegetation coverage is a crucial ecological indicator, displaying the amount of lush vegetation and the photosynthetic area (Baumgartner & Cherlet, 2015). At diverse spatial and temporal scales, different plant species respond to hydrothermal variables in different ways. Additionally, grassland NEP significantly increases with increased precipitation (Zhang et al., 2014). During dry years, grasses become a net carbon source, particularly in arid and semi-arid grasslands. There is a strong temporal-spatial relationship between precipitation patterns and grassland production (Barnett & Facey, 2016). They aid in the preservation of hydrological cycles by facilitating water infiltration into the soil (Döbert et al., 2021).

At the moment, methods for field observation and inspection are mostly used to monitor the degradation of grasslands. The biological indicators of grassland deterioration are primarily soil characteristics and biological indicators (such as vegetation cover, biomass, dominating species, etc. (Zomer & Ramsay, 2018). Regional scales of soil attributes are very difficult to collect, although large-scale grassland deterioration is frequently assessed by using biological markers. At large regional scales, information on vegetation cover, biomass, etc., can be obtained via remote sensing techniques (Ali et al., 2016). The annual variations in grassland biomass could be a result of climatic changes or the deterioration of the grassland itself (Fitton et al., 2017). Thus, there are many factors responsible for grassland degradation. Some of the factors are discussed here.

3.1. Conversion and Fragmentation

Urbanization and the conversion of grassland to farmland have a negative influence on soil resources because they result in the loss of up to 90% of the soil structure that is necessary for infiltration and water storage in soils. There is a reduction in pollination functions, soil organic carbons, and drainage changes in hydrology (Nitsch et al., 2012).

3.2. Grazing

Excessive grazing can cause profitable forage plants to be replaced by species with inferior forage quality, increasing the cover of bare soil while the contribution of very high-quality grass species declines (Li et al., 2022). As soil bulk density rises, it can harm ecosystem processes like water penetration, changes to biotic and abiotic traits, and resilience capacity reduce (Liu et al., 2019).

Overgrazing happens when plants are subjected to intense grazing pressure for prolonged periods or without enough time to recover. Overgrazing which is often regarded as the leading cause of desertification and erosion decreases the utility, production, plant vigor, and biodiversity of the land. Additionally, it may cause the proliferation of weeds and other non-native plant species (Mbaabu et al., 2020). From the non-degraded to the degraded patch, bacterial richness increased while fungal richness declined. The relative abundance of rare soil microbial species could be a helpful ecological indicator for the degradation of grasslands (Luo et al., 2023).

Su et al., (2005) examined characteristics of vegetation and soil properties under continuous grazing and exclusion of livestock for 5 and 10 years in representative degraded sandy grassland. Excluding livestock grazing enhances vegetation recovery, litter accumulation, and the development of annual and perennial grasses. Soil organic carbon and total N concentrations, soil biological properties including some enzyme activities and basal soil respiration improved following a 10-year exclusion of livestock, suggesting that degradation of the grassland is being reversed.

3.3. Construction of Roads

Roads divide formerly interconnected ecosystems into distinct, ecologically more fragile portions. Numerous grassland bird species‘ regular mating activities are greatly reduced by road activities. Highways seem to harm the degradation of good quality grassland resources. Grasslands are the only indication for the restoration and recovery of wastelands (Reijnen et al., 1996).

3.4. Effects of Degradation on Sustainable Agriculture

Measures to control cropland erosion should be introduced, including the construction of windbreaks, stubble cultivation, increased organic fertilizer, and spring irrigation (Nan, 2005).

3.5. Fire

Fire has been illegal under state law because it leads to reducing humus, biomass, and extinction of many grassland species in the world. The land has a very specific feature it is capable of resprouting the grass from subsurface organs of soil layers (Zomer & Ramsay, 2018). The grasslands are related to repeated fires that do not permit the forest to grow. The grasses are the major producers of biomass in the given regions. Each grassland ecosystem has a wide variety of species of grasses and herbs. Some grass and herb species are more sensitive to excessive grazing and are suppressed if the area is overgrazed. Others are destroyed by repeated fires and cannot regenerate. Thus, overused or frequently burnt grasslands are degraded and are poor in plant species diversity. Besides these, there are many other anthropogenic factors responsible for grassland degradation.

There are various drivers, threats and consequences of grassland degradation (Figure 3). It causes:

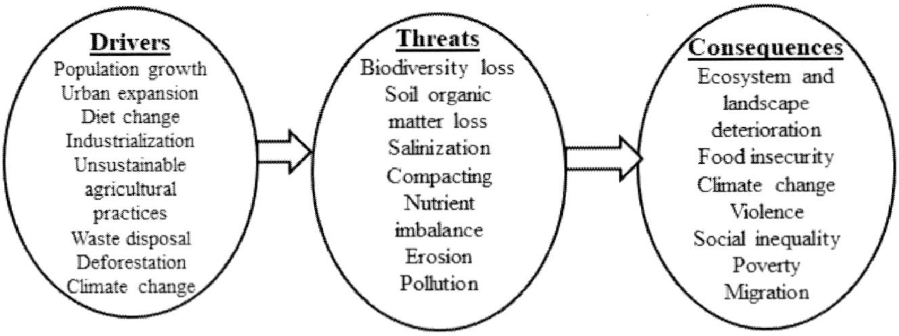

Figure 3. Drivers, threats and consequences of grassland degradation.

3.6. Increases in GHG Emissions

Rotational grazing with a moderate intensity improves soil fertility, plant growth, and carbon and nitrogen storage while lowering GHG emissions.

3.7. Conflict with Other Practice(s)

Urbanization is the main cause of the elimination of grasslands in this present scenario (Gang et al., 2014).

3.8. Other Conflicts

The restoration of deteriorated grasslands might be hampered by land ownership and funding.

4. Grassland Restoration

The primary function of grasslands is to provide free-range grazing to domestic animals and other wildlife. Grasslands restoration techniques vary as per intended major use. We can restore grasslands by using ecological approaches like protection from grazing, soil and water conservation, controlled fire, bush clearing, application of fertilizers or by using grasses and legumes.

- Fencing for the protection of grazing
- Clearing
- Bush control
- Controlled fire
- Soil and water conservation

Fencing and nitrogen fertilization are typical and successful restoration techniques for grasslands (Akiyama & Kawamura, 2007). The operation of ecosystems is heavily regulated by soil microorganisms. However, we currently lack knowledge on how below-ground microbial communities are affected by grassland degradation.

Table 1. Incentives for the protection and restoration of grasslands

Incentives for the protection of grasslands	
Regulatory incentives	National or sub-national legislation, e.g., land use and agricultural zonation laws, nature protection and environment protection laws Grassland and grazing laws Prohibitions on habitat destruction/vegetation conversion Land use planning
Private law mechanisms	Easements, covenants Private contracts for conservation
Financial incentives	Taxation policies, e.g., related to easements
Incentives for improved management of grasslands	
Regulatory incentives	National or sub-national legislation, e.g., land use and agricultural zonation laws, nature protection and environment protection laws Grassland and grazing laws
Financial aid	Credit policies
Grant aid	Government cost-sharing grants (e.g., Landcare grants) Input subsidies Technical assistance (extension services)
Conditional payments	Payments for specified practices Payments for specified ecosystem services Input subsidies Off-take subsidies One-off grant payments Recurring payments for ecosystem services
Product market payments	Labeling for niche products (e.g., Geographical Indications) Certification of products Eco-tourism revenues

Accumulation of soil organic matter, soil fertility, and vegetation development is closely correlated with grassland restoration as reported by Carbutt & Kirkman (2022). To diagnose the stages of recovery from degraded grasslands, not only by the life form and the growth form of dominant species, but also the measurements of vegetation such as coverage, plant height, number of annual and perennial species, and the existence of degradation indicator species are effective. Agro-ecological countermeasures against fixing moving dunes and rehabilitation in the sandy land. There are many benefits of grassland restoration; some are listed in Table 1.

The usage of sheep dung as a water retention agent to establish a method for low input improvement of degraded grassland in a semiarid area (Deng et al., 2011). The germination from dung pellets showed a good result when pellets were embedded 1 cm into the ground. The dung pellets were buried in the wet soil, those pellets were capable of absorbing approximately 2.2-fold their mass in water and retaining it for a few days.

Biodiversity conservation and livestock production can be seen as complementary management goals, allowing for sustainable use. Fire and grazing are selective forces that change the composition of grasslands. Management is required for the maintenance of diverse and productive grasslands in this region. There should be revisions in subsidies and incentives.

4.1. Introduction of Alien C3 and C4 Species and Fertilization for Restoration

To improve fodder quality and quantity, particularly during the winter natural grasslands are over-sown with alien species (Malaviya et al., 2021). This practice is frequently paired with fertilizing and liming. While many introduced species, such as Lolium multiflorum and certain European Fabaceae family species have higher production and nutritional potential than native species which are mostly available in the seed market (Alonso et al., 1999). Ultimately there will be some threats from these alien species. Using a mixture of both native and non-native forage species sowing can bring at least partially control the populations of the invasive species (Ruan et al., 2021).

4.2. Potential for Self-Recovery and Restoration after Management-Induced Degradation

Self-recovery may be more challenging or even impossible when soil conditions have been altered, such as by fertilization or liming. Prescribed fire may be a conservation technique in regions where grazing is impractical such as in conservation units (Sun et al., 2022).

The amount of nutrients in the soil is raised by using arable lands, it is generally recognized that when grasslands are plowed and turned into arable fields, significant amounts of carbon are stored which can be easily lost into the atmosphere. No-tillage methods provide smaller losses of carbon in soils than conventional tilling (Kumar et al., 2012).

5. Benefits of Restoration of Degraded Grasslands

5.1. Improvement of Soil Properties

Soils maintain C and N ratios on a long-term basis. legume species in the grassland even be more advantageous for nitrogen fixing in soils and improving the cycle of nutrients. Pollutants and contaminants might be decreased or eliminated (He et al., 2015). Increased soil C content may aid in reducing compaction. Water use efficiency may be improved by improved soil structure (Guo et al., 2021).

5.2. Increases in Production (e.g., Food/Fuel/Feed/Timber/Fiber)

To feed cattle, restored grasslands will yield more and better (Shen et al., 2020).

5.3. Mitigation and Adaptation to Climate Change

Restored grassland could offset N2O and CH4 emissions through C sequestration while adapting to changing climatic conditions (Sheng et al., 2022).

5.4. Socio-Economic Benefits

Through small-scale farming and pastoralism, grasslands serve as a significant source of revenue in developing regions. Restoration of grasslands increases the income of rural communities and enterprises. Food security and the production of animals on grasslands are closely related (Wang et al., 2014). Improved grassland or pasture management can reduce the danger of land degradation in this socioeconomic setting and aid in the restoration of degraded lands, resulting in better plant and litter cover, decreased erosive vulnerability, reduced soil compaction, and increased biodiversity (Zhou et al., 2020).

5.5. Other Benefits

The aesthetic benefits of a landscape enable tourism and improve human well-being. To prevent the deterioration of grasslands, mixed production systems provide additional feed supplements for livestock products and resource management (Bengtsson et al., 2019).

Conclusion

Grasslands play an important role in livestock farming and environmental conservation. According to remote sensing and GIS data, 49% of the total grasslands are degraded globally as a result of intensive agriculture and land conversion since the 1940s. Grassland degradation can help to inform policy and practice about the level of grassland degradation in a certain region. The supply of multiple ecosystem services fall short and becomes a demand by grassland stakeholders. Ongoing climate change also poses a threat to all grasslands, as it causes grassland degradation. Projected future climate change will likely combine with human activities to cause increased woody plant encroachment in some areas and desertification in others. Increases in temperature and precipitation both encourage the growth of grass, but different types of grasslands respond to climatic conditions in different ways. Degradation and restoration of grasslands coexist at the same time, with the area of degraded grasslands being somewhat smaller than that of restored grasslands. For the restoration of grassland, we can use ecological, biological, or biotechnological tools. But the best one is integrated techniques of grassland management.

References

Abdalla, K., Mutema, M., Chivenge, P., Everson, C., & Chaplot, V. (2018). Grassland degradation significantly enhances soil CO_2 emission. *CATENA*, *167*, 284–292. https://doi.org/10.1016/j.catena.2018.05.010.

Akiyama, T., & Kawamura, K. (2007). Grassland degradation in China: Methods of monitoring, management, and restoration. *Grassland Science*, *53*(1), 1–17. https://doi.org/10.1111/j.1744-697X.2007.00073.x.

Ali, I., Cawkwell, F., Dwyer, E., Barrett, B., & Green, S. (2016). Satellite remote sensing of grasslands: From observation to management. *Journal of Plant Ecology*, *9*(6), 649–671. https://doi.org/10.1093/jpe/rtw005.

Alonso, S. I., Guma, I. R., & Clausen, A. M. (1999). Variability for salt tolerance during germination in Lolium multiflorum Lam. naturalized in the pampean grasslands. *Genetic Resources and Crop Evolution*, *46*(1), 87–94. https://doi.org/10.1023/A:1008638325484.

Andrade, B. O., Koch, C., Boldrini, I. I., Vélez-Martin, E., Hasenack, H., Hermann, J. M., Kollmann, J., Pillar, V. D., & Overbeck, G. E. (2015). Grassland degradation and restoration: A conceptual framework of stages and thresholds illustrated by southern Brazilian grasslands. *Natureza and Conservação*, *13*(2), 95–104. https://doi.org/10.1016/j.ncon.2015.08.002.

Bardgett, R. D., Bullock, J. M., Lavorel, S., Manning, P., Schaffner, U., Ostle, N., Chomel, M., Durigan, G., L. Fry, E., Johnson, D., Lavallee, J. M., Le Provost, G., Luo, S., Png, K., Sankaran, M., Hou, X., Zhou, H., Ma, L., Shi, H. (2021). Combatting global grassland degradation. *Nature Reviews Earth and Environment*, *2*(10), 720–735. https://doi.org/10.1038/s43017-021-00207-2.

Barnett, K. L., & Facey, S. L. (2016). Grasslands, invertebrates, and precipitation: A review of the effects of climate change. *Frontiers in Plant Science*, *7*, (AUG2016), 1–8. https://doi.org/10.3389/fpls.2016.01196.

Baumgartner, P., & Cherlet, J. (2015). Institutional framework of (in) action against land degradation. In *Economics of Land Degradation and Improvement - à Global Assessment for Sustainable Development*. https://doi.org/10.1007/978-3-319-19168-3_3.

Bengtsson, J., Bullock, J. M., Egoh, B., Everson, C., Everson, T., O'Connor, T., O'Farrell, P. J., Smith, H. G., & Lindborg, R. (2019). Grasslands—More important for ecosystem services than you might think. *Ecosphere*, *10*(2). https://doi.org/10.1002/ecs2.2582.

Breidenbach, A., Schleuss, P. M., Liu, S., Schneider, D., Dippold, M. A., de la Haye, T., Miehe, G., Heitkamp, F., Seeber, E., Mason-Jones, K., Xu, X., Huanming, Y., Xu, J., Dorji, T., Gube, M., Norf, H., Meier, J., Guggenberger, G., Kuzyakov, Y., & Spielvogel, S. (2022). Microbial functional changes mark the irreversible course of Tibetan grassland degradation. *Nature Communications*, *13*(1), 1–10. https://doi.org/10.1038/s41467-022-30047-7.

Bruinenberg, M. H., Valk, H., Korevaar, H., & Struik, P. C. (2002). Factors affecting digestibility of temperate forages from seminatural grasslands: A review. *Grass and Forage Science*, *57*(3), 292–301. https://doi.org/10.1046/j.1365-2494.2002.00327.x.

Carbutt, C., & Kirkman, K. (2022). Ecological grassland restoration—A South African perspective. *Land*, *11*(4), 1–25. https://doi.org/10.3390/land11040575.

Deng, X., Huang, J., Huang, Q., Rozelle, S., & Gibson, J. (2011). Do roads lead to grassland degradation or restoration? A case study in Inner Mongolia, China. *Environment and Development Economics*, *16*(6), 751–773. https://doi.org/10.1017/S1355770X11000180.

Döbert, T. F., Bork, E. W., Apfelbaum, S., Carlyle, C. N., Chang, S. X., Khatri-Chhetri, U., Silva Sobrinho, L., Thompson, R., & Boyce, M. S. (2021). Adaptive multi-paddock grazing improves water infiltration in Canadian grassland soils. *Geoderma*, *401*(June). https://doi.org/10.1016/j.geoderma.2021.115314.

Fitton, N., Datta, A., Cloy, J. M., Rees, R. M., Topp, C. F. E., Bell, M. J., Cardenas, L. M., Williams, J., Smith, K., Thorman, R., Watson, C. J., McGeough, K. L., Kuhnert, M., Hastings, A., Anthony, S., Chadwick, D., & Smith, P. (2017). Modeling spatial and interannual variations of nitrous oxide emissions from UK cropland and grasslands using DailyDayCent. *Agriculture, Ecosystems and Environment, 250*(August), 1–11. https://doi.org/10.1016/j.agee.2017.08.032.

Fondell, T. F., & Ball, I. J. (2004). Density and success of bird nests relative to grazing on western Montana grasslands. *Biological Conservation, 117*(2), 203–213. https://doi.org/10.1016/S0006-3207(03)00293-3.

Gang, C., Zhou, W., Chen, Y., Wang, Z., Sun, Z., Li, J., Qi, J., & Odeh, I. (2014). Quantitative assessment of the contributions of climate change and human activities on global grassland degradation. *Environmental Earth Sciences, 72*(11), 4273–4282. https://doi.org/10.1007/s12665-014-3322-6.

Guo, X., Zhou, H., Dai, L., Li, J., Zhang, F., Li, Y., Lin, L., Li, Q., Qian, D., Fan, B., Lan, Y., Si, M., Li, B., Cao, G., Du, Y., & Wang, B. (2021). Restoration of degraded grassland significantly improves water storage in alpine grasslands in the Qinghai-Tibet Plateau. *Frontiers in Plant Science, 12*(December), 778656. https://doi.org/10.3389/fpls.2021.778656.

He, C., Tian, J., Gao, B., & Zhao, Y. (2015). Differentiating climate- and human-induced drivers of grassland degradation in the Liao River Basin, China. *Environmental Monitoring and Assessment, 187*(1), 4199. https://doi.org/10.1007/s10661-014-4199-2.

Jiang, G., Han, X., & Wu, J. (2006). Restoration and management of the Inner Mongolia grassland require a sustainable strategy. *AMBIO: A Journal of the Human Environment, 35*(5), 269–270. https://doi.org/10.1579/06-s-158.1.

Kumar, B. M., Singh, A. K., & Dhyani, S. K. (2012). *South Asian agroforestry: Traditions, transformations, and prospects (issue July)*. https://doi.org/10.1007/978-94-007-4676-3_19.

Lauenroth, W. K., & Adler, P. B. (2008). Demography of perennial grassland plants: Survival, life expectancy, and life span. *Journal of Ecology, 96*(5), 1023–1032. https://doi.org/10.1111/j.1365-2745.2008.01415.x.

Li, M., Liu, S., Wang, F., Liu, H., Liu, Y., & Wang, Q. (2022). Cost–benefit analysis of ecological restoration based on land use scenario simulation and ecosystem service on the Qinghai-Tibet Plateau. *Global Ecology and Conservation, 34*, e02006. https://doi.org/10.1016/j.gecco.2022.e02006.

Liu, Y., Zhang, Z., Tong, L., Khalifa, M., Wang, Q., Gang, C., Wang, Z., Li, J., & Sun, Z. (2019). Assessing the effects of climate variation and human activities on grassland degradation and restoration across the globe. *Ecological Indicators, 106*, 105504. https://doi.org/10.1016/j.ecolind.2019.105504.

Luo, S., Png, G. K., Ostle, N. J., Zhou, H., Hou, X., Luo, C., Quinton, J. N., Schaffner, U., Sweeney, C., Wang, D., Wu, J., Wu, Y., & Bardgett, R. D. (2023). Grassland degradation-induced declines in soil fungal complexity reduce fungal community stability and ecosystem multifunctionality. *Soil Biology and Biochemistry, 176*, 108865. https://doi.org/10.1016/j.soilbio.2022.108865.

Malaviya, D. R., Roy, A. K., & Kaushal, P. (2021). *UKnowledge potential of C4 tropical grasses to contribute in carbon sequestration, environmental security and livelihood opportunities through increased fodder availability.*

Mbaabu, P. R., Olago, D., Gichaba, M., Eckert, S., Eschen, R., Oriaso, S., Choge, S. K., Linders, T. E. W., & Schaffner, U. (2020). Restoration of degraded grasslands, but not an invasion by Prosopis juliflora, avoids trade-offs between climate change mitigation and other ecosystem services. *Scientific Reports*, *10*(1), 1–13. https://doi.org/10.1038/s41598-020-77126-7.

Nan, Z. (2005). The grassland farming system and sustainable agricultural development in China. *Grassland Science*, *51*(1), 15–19. https://doi.org/10.1111/j.1744-697X.2005.00003.x.

Nitsch, H., Osterburg, B., Roggendorf, W., & Laggner, B. (2012). Cross compliance and the protection of grassland – Illustrative analyses of land use transitions between permanent grassland and arable land in German regions. *Land Use Policy*, *29*(2), 440–448. https://doi.org/10.1016/j.landusepol.2011.09.001.

Reijnen, R., Foppen, R., & Meeuwsen, H. (1996). The effects of traffic on the density of breeding birds in Dutch agricultural grasslands. *Biological Conservation*, *75*(3), 255–260. https://doi.org/10.1016/0006-3207(95)00074-7.

Ruan, H., Wu, X., Wang, S., Yang, J., Zhu, H., Guo, Q., Wang, L., & Wang, D. (2021). The responses of different insect guilds to grassland degradation in northeastern China. *Ecological Indicators*, *133*, 108369. https://doi.org/10.1016/j.ecolind.2021.108369.

Shen, X., Yang, F., Xiao, C., & Zhou, Y. (2020). Increased contribution of root exudates to soil carbon input during grassland degradation. *Soil Biology and Biochemistry*, *146*(January), 107817. https://doi.org/10.1016/j.soilbio.2020.107817.

Sheng, J., Zhou, M., Guo, Y., Yuan, Y., Li, X., Zhang, W. H., & Bai, W. (2022). Aboveground productivity and community stability tend to keep stable under long-term fencing and nitrogen fertilization on the restoration of degraded grassland. *Ecological Indicators*, *140*(May), 108971. https://doi.org/10.1016/j.ecolind.2022.108971.

Su, Y. Z., Li Y. L., Cui J. Y., & Zhao W. Z. (2005). Influences of continuous grazing and livestock exclusion on soil properties in a degraded sandy grassland, Inner Mongolia, northern China Catena, 59, 267 -278.

Sun, J., Wang, Y., Piao, S., Liu, M., Han, G., Li, J., Liang, E., Lee, T. M., Liu, G., Wilkes, A., Liu, S., Zhao, W., Zhou, H., Yibeltal, M., Berihun, M. L., Browning, D., Fenta, A. A., Tsunekawa, A., Brown, J., Tsubo, M. (2022). Toward a sustainable grassland ecosystem worldwide. *Innovation*, *3*(4), 100265. https://doi.org/10.1016/j.xinn.2022.100265.

Wang, X., Dong, S., Yang, B., Li, Y., & Su, X. (2014). The effects of grassland degradation on plant diversity, primary productivity, and soil fertility in the alpine region of Asia's headwaters. *Environmental Monitoring and Assessment*, *186*(10), 6903–6917. https://doi.org/10.1007/s10661-014-3898-z.

Wang, Z., Deng, X., Song, W., Li, Z., & Chen, J. (2017). What is the main cause of grassland degradation? A case study of grassland ecosystem service in the middle-south Inner Mongolia. *CATENA*, *150*, 100–107. https://doi.org/10.1016/j.catena.2016.11.014.

When, S., Hovstad, K.A., Johansen, L. (2018). The relationships between biodiversity and ecosystem services and the effects of grazing cessation in semi-natural grasslands. *Web Ecology* 18(1):55-65.

Winfree, R., Fox, J.W., Williams, N.M., Reilly, J.R., Cariveau, D.P. (2015). Abundance of common species, not species richness, drives delivery of a real-world ecosystem service. *Ecology Letters* 18(7):626-635.

Yan, Y., Xu, X., Xin, X., Yang, G., Wang, X., Yan, R., Chen, B. (2011). Effect of vegetation coverage on aeolian dust accumulation in a semiarid steppe of northern China. *Catena* 7(3): 351-356.

Zhang, B., Wang, Y., Li, J., & Zheng, L. (2022). Degradation or restoration? The temporal-spatial evolution of. *Land*.

Zhang, L., Guo, H., Jia, G., Wylie, B., Gilmanov, T., Howard, D., Ji, L., Xiao, J., Li, J., Yuan, W., Zhao, T., Chen, S., Zhou, G., & Kato, T. (2014). Net ecosystem productivity of temperate grasslands in northern China: An upscaling study. *Agricultural and Forest Meteorology*, *184*, 71–81. https://doi.org/10.1016/j.agrformet.2013.09.004.

Zhang, X., Niu, J., Buyantuev, A., Zhang, Q., Dong, J., Kang, S., Zhang, J. (2016) Understanding grassland degradation and restoration from the perspective of ecosystem services: A case study of the Xilin River Basin in Inner Mongolia, China. *Sustainability* 8(7):594-603.

Zhou, W., Li, J., & Yue, T. (2020). Springer geography remote sensing monitoring and evaluation of degraded grassland in China accounting of grassland carbon source and carbon sink. http://www.springer.com/series/10180.

Zomer, M. A., & Ramsay, P. M. (2018). Espeletia giant rosette plants are reliable biological indicators of time since fire in Andean grasslands. *Plant Ecology*, *219*(1), 79–88. https://doi.org/10.1007/s11258-017-0779-x.

Chapter 11

The Restoration of Wastelands: Methods of Landfill Reclamation

Vijai Krishna*
Department of Environmental Sciences, (RGSC), Banaras Hindu University, Varanasi, India

Abstract

Any land not suitable for any use is commonly known as wasteland. A large part of India comes under different categories of wasteland. These classifications generally do not include the landfill as a wasteland but due to the inherent nature of landfills, these areas also act like a wasteland. This paper is about the understanding of the description of basic characteristics, spatial extent and classification of wasteland so that to understand that the landfills should also be considered as wasteland. Landfill creates several types of environmental as well as socio-economic problems too. The leachate created within landfill contaminates the underground water resources. The gas generated within landfill pollutes the nearby atmosphere. The landfill also acts as a breeding place for vectors of infectious diseases. Landfill fire, landfill slide, and landfill subsidence are also great hazards. Overall, this paper summarizes the methods of waste management, types of landfills and the methods of landfill reclamation with its benefits and drawbacks. Landfill reclamation can transform a wasteland into a potentially useful place for commercial or recreational activities. Major methods of landfill reclamation include photocopying, landfill mining, in-situ aeration, Landfill surcharging, Soil vapor extraction and air sparging, treatment of landfill leachate, etc.

* Corresponding Author's Email: vijaikrishna@bhu.ac.in.

In: Conserving, Halting and Reinstating Degraded Natural Resources …
Editors: Rajani Srivastava and Pankaj Kumar
ISBN: 979-8-89113-067-8

Keywords: wasteland, landfill, solid waste, phytocap, landfill mining, *in situ* aeration

1. Introduction

Agarwal (1986) defines wasteland as lands producing less than its possibilities and productivity which could be amplified by rational efforts and venture. Bhumla & Khare (1984) defined wastelands as those which are (a) ecologically unsteady (b) topsoil has been approximately lost and (c) developed toxicity in the root zones. National Remote Sensing Agency (NRSA, 1991) described wasteland as degraded land which can be transformed into vegetative protection with sensible efforts and which is worsening due to deficiency of water and soil management or due to certain natural reasons. Wastelands may be the result of incapacities such as location, environment, chemical and physical characteristics of the soil, or financial or management restrictions. Stamp (1948) defined wasteland as the land which has been abandoned and for which no further use has been found. Wasteland Survey and Reclamation Committee (1961) defined wastelands are either not accessible or left out of agriculture as fallows and cultivable waste as wastelands. According to National Wasteland Development Board (NWDB, 1986), wastelands are defined as that land is degraded and is currently unutilized except as currently fallow due to diverse restrictions. Those lands are currently unutilized or are not being used to their finest potential due to limitations (Sinha & Dubey, 1986). Shafi (1968) defined wasteland as uncultivated land excluding fallows land. Ranade (2009) suggested that Wastelands are those areas that are not used to the level of their optimum potential but can be brought under vegetation cover with some effort.

2. Classification of Wastelands

According to National Wasteland Development Board (NWDB, 1985), wastelands are defined as those lands which are degraded and are presently lying unutilized except as currently fallow due to different constraints. The wastelands have been classified as follows:

(I) Culturable wastelands, and
(II) Unculturable wastelands.

Table 1. Classification of wastelands (NWDB, 1985)

S. N.	Classes	Example
(A)	Culturable Wastelands:	
(1)	Wastelands that are unutilized, partially utilized, or mismanaged (State occupation or notified forest Areas)	(a) Gullied and or ravenous land (b) Undulating upland with or without scrubs (c) Surface waterlogged land and marsh areas (d) Salt-affected land
(2)	Category of wastelands based on Ecological limitations	(e) Shifting cultivation areas (f) Degraded forest land (g) Degraded pastures/grazing land (h) Degraded non-forest plantation land (i) Strip lands (j) Sands (k) Mining spoils/industrial wastelands.
(B)	Unculturable Wastelands:	
(1)	Category of wastelands which are more or less unavailable	(a) Barren rocky, stony wastes, sheetrock Area (b) Steep sloping area (c) Snow-covered and or glacial area

2.1. Estimation of the Wasteland

In 2015–16, 5,57,665.51 sq. km. of wastelands were recorded nationwide (almost 17% of India‘s geographical area) compared to 5,66,070.36 sq. km. in 2008–09. This shows that between 2008–09 and 2015–16, 8,404.86 sq. km. of the nation’s various wastelands classifications were converted to non–wastelands (Wastelands Atlas of India, 2019).

3. Municipal Solid Waste Management (MSW)

The term “MSW” refers to trash produced by people, institutions, businesses, municipal facilities, and treatment facilities (Tchobanoglous et al., 1993). Together, degradable and non-degradable materials such as leather, plastics, rubbers, metals, glass, ash from fuel burning such as coal, briquettes, or woods, and dust make up municipal solid waste (MSW). Degradable materials include food waste, straw and yard waste, paper, textiles, and partially degradable materials such as wood, disposable napkins, and sludge

(Gupta et al., 2015). With population increase, which is inextricably linked to industrialization, urbanization, and economic development, MSW generation is gradually rising. As a result, the widespread dumping of MSW is steadily getting worse due to a lack of social and environmental awareness. The main factors affecting MSW's composition include geographic locations, dietary preferences, climatic circumstances, degree of income, social and cultural norms, and the presence of rules and restrictions (Gupta et al., 2015; Late and Mule, 2013; Yadav and Devi, 2009). The appropriate management of MSW is currently the most urgent demand since there is a severe lack of land available for the safe disposal of generated solid waste. For proper waste segregation, a variety of approaches including density separation, electric field separation, and magnetic field separation have been used. However, different waste processing techniques have been used around the world as a result of technological advancements. These techniques include aerobic composting, vermicomposting, anaerobic digestion, incineration, pyrolysis, gasification, pelletization, etc. (Karmakar et al., 2014). Waste processing, also known as waste conversion, is the process of turning waste materials into usable energy in the form of heat, electricity, or fuel (Dutta, 2015). Both thermal processing and bioprocessing are included.

3.1. Bioprocessing

Bioprocessing is the process of transforming waste materials into useful products using living organisms. It includes anaerobic digesting, aerobic composting, and vermicomposting, among other things. One of the less expensive methods of treating MSW's organic content is aerobic composting (Gupta et al., 2015). In a matter of four to six weeks, the biological decomposition of organic materials by microorganisms in an aerobic environment under hot, humid conditions results in the stable organic component known as compost (Tchobanoglous et al., 1993). For the microbes to stabilize the organic matter, they need 30 parts of carbon for every 1 part of nitrogen in the raw material needed for aerobic composting (CPHEEO, 2000). This type of processing or decomposition is best for the lignocellulosic-rich garden or yard trash as well as the percentage of MSW that is organically biodegradable. In the aerobic composting process, aerobic bacteria quickly oxidize organic materials to produce stable organic matter (compost), carbon dioxide, ammonia, sulfate, and water. Compost is a soil stabilizer that contains micronutrients necessary for plant growth as well as plant nutrients (nitrogen, phosphorus, and potassium). Additionally, compost and chemical fertilizers together produce the best outcomes for plant growth (Gautam et al., 2010).

3.2. Vermicomposting

Vermicomposting is a process in which the earthworms transform the organic fraction of MSW into nutrient-rich soil. Vermicomposting includes the integrated action of earthworms and microorganisms. It involves primary decomposition and secondary decomposition. Primary decomposition indicates the microbial decomposition of biodegradable organic matter through extracellular enzymatic activities whereas secondary decomposition denotes decomposition by microorganisms inhabiting the alimentary tract of earthworms (Ndegwa and Thompson, 2001). Vermicomposting is a three-factor system that entails microorganisms, earthworms and biomass and depends on abiotic factors like temperature, moisture, aeration, etc. The optimum conditions required for the earthworms to thrive are a moisture content of 20-80%, a temperature of 20-40°C and 40% of fermentable organic substrate. The presence of high concentrations of heavy metals like Cd, Cr, Pb and Zn has ill effects on the growth of earthworms. However, nowadays the practice of vermicomposting is increasing on segregated MSW.

3.3. Anaerobic Digestion

Anaerobic digestion is a process in which the biodegradable organic portion of MSW decomposed by microorganisms in the absence of air to a combination of carbon dioxide and methane (Biogas) (Holm-Nielsen et al., 2009). The other substrates which are produced in association with anaerobic digestion are ammonia, hydrogen sulfide and resistant organic matter. Biogas can be highly efficient as a fuel or can be used to generate on-site electricity (Annepu, 2012). Resistant organic matter can also be utilized as manure for agriculture after its process of stabilization.

3.4. Thermal Processing

Using thermal energy, thermal processing converts waste materials into usable products. It entails incineration, pelletization, gasification, and pyrolysis. Incineration includes the burning of unprocessed MSW at a temperature of between 800 and 1000°C in extra air resulting in the production of bottom ash, energy, and gases such as carbon dioxide, oxygen, nitrogen oxides, sulfur dioxide, and water (Zaman, 2009). Toxic dioxin and furan production must be stopped by maintaining these high temperatures. There is always a way to use heat energy to produce steam, which steam engines subsequently use to generate electricity. The trash must have a moisture content of less than 45% and a calorific value of more than 1500 Kcal/Kg as minimum conditions for successful use of incineration (Rajput et al., 2009). Pyrolysis is a method of thermal decomposition or destructive distillation or carbonization of organic waste material by maintaining an average temperature of 900°C in an oxygen-free environment, This results in the production of a mixture of combustible (carbon monoxide, methane, hydrogen, and ethane) and noncombustible gases (carbon dioxide, water, nitrogen), a pyroligenous liquid (tar, pitch, light oil, and low boiling chemicals like acetic acid, acetone, methanol) and charcoal. The three pyrolysis byproducts have a high calorific value and can be effectively used in various combustion processes. Additionally, the liquid works well as fuel oil (CPHEEO, 2000). In this technology, producing high temperatures requires a significant quantity of electrical energy. The chemical composition of organic matter and the operating conditions (pyrolysis temperature, heating time, pressure, etc.) completely determine the number of final products (Yang et al., 2007). In contrast, gasification is a technique for thermally decomposing organic matter at extremely high temperatures (greater than 1000°C) in a setting with limited airflow, producing a mix of combustible and non-combustible gases. The production of high calorific value gaseous products at these high temperatures eventually results in the generation of electrical power. Typically, wood or pulp materials are decomposed via pyrolysis or gasification. However, both approaches are becoming more used nowadays for the treatment of MSW (CPHEEO, 2000). Additionally, industrialized nations can use gasification to generate electricity, which lowers greenhouse gas emissions by displacing the usage of fossil fuels. Gasification is another economically viable option for developing nations that can supply electricity to rural areas using locally accessible biomass waste (McKendry, 2002). When waste materials are turned into fuel pellets through the procedures of drying, separating combustibles, and reducing waste matter in conjunction with various additives, the process is known as pelletization. The final products are referred to as refuse-derived fuel (RDF), which is widely used as an input fuel feed for industrial furnaces or incinerating technologies. Plastic and other organic materials, as well as biodegradable elements, are used to prepare RDF (Gilbert et al., 2009). To reach the appropriate calorific values, MSW must be mechanically dried out or simply sun dried if its moisture content is greater than 35–40%. Consequently, the process of pelletization to create RDF can be an appropriate waste management choice to reduce MSW (CPHEEO, 2000).

3.5. Landfilling

Due to environmental concerns, all of the waste processing technologies mentioned above are widely employed throughout the world. However, since all waste processing procedures leave behind some garbage, landfilling cannot be avoided. Additionally, roughly 30% of the MSW produced in India is made up of inert elements that must also be dumped at a designed landfill. Due to financial restrictions, open dumping or unregulated landfilling is the most widely used method to manage MSW in underdeveloped nations. Leachate or gas collection systems are absent from uncontrolled landfills. Additionally, uncontrolled landfilling is more cost-effective to carry out and better suited for mixed-waste types. The practice of landfilling is the most straightforward and economical way to manage MSW. For garbage of mixed types that cannot be treated by existing waste processing systems, landfills are used. Additionally, non-hazardous wastes or waste products from pre- or post-processing methods can be properly dumped in landfills (CPHEEO, 2000). With a large proportion of organic elements in MSW, landfills are linked to leachate formation and gas emissions. Therefore, geo-membrane liners should be installed in landfills to stop leachate from mixing with the aquifer systems that are connected to them as well as the nearby surface water bodies. Along with the arrangements for daily cover, landfills should also be equipped with leachate and gas collection systems.

4. Types of Landfills

The principal methods used for landfilling may be classified (Tchobanoglous et al., 1993) as:

4.1. Area Method

The area method is used when the terrain is unsuitable for the excavation of trenches in which to place the solid wastes. The filling operation usually is started by building an earthen level against which waste is placed in thin layers and compacted. Each layer is compacted as the filling progresses until the thickness of the compacted waste reaches a height varying from 2 to 3m (6 to 10 ft). At that time, and the end of each day's operation, a 150 to 300 mm (6-12 in) layer of cover material is placed over the completed fill. The cover material must be hauled in by truck or earth-moving equipment from adjacent land or from borrowed pit areas. In some newer landfill operations, the daily cover material is omitted. A completed lift, including the cover material, is called a cell. Successive lifts are placed on top of one another until the final grade called for in the ultimate development plan is reached. The final layer of cover material is used when the fill reaches the final design height.

4.2. Trench Method

This landfill is ideally suited to areas where an adequate depth of cover material is available at the site and where the water table is well below the surface. To start the process (for a small landfill), a portion of the trench is dug with a bulldozer and the dirt is stockpiled to form an embankment behind the first trench. Wastes are then placed in the trench, spread into thin layers and compacted. The operation continues until the desired height is reached. Cover material is obtained by excavating an adjacent trench or continuing the trench that is being filled, in large landfills, a dragline and one or more scrapes are used to excavate a deep rectangular pit.

4.3. Depression Landfills

At locations where natural or artificial depression exists, it is often possible to use them efficiently for landfilling operations. Canyons, ravines, dry borrow pits, and quarries have all been used for this purpose. The techniques to place and compact solid wastes in depression landfills vary with the geometry of the site, the characteristics of the cover material, the hydrology and geology of the site, and access to the site. In a canyon site, filling starts at the head end of the canyon and ends at the mouth. This practice prevents the accumulation of water behind the landfill. Wastes usually are deposited on the canyon floor and from there are pushed up against the canyon face at a slope of about 2 to 1. In this way, a high degree of compaction can be achieved.

According to Narayana (2009), there are three different sorts of landfills:

I. *Uncontrolled Landfills or Open Dump Sites* - Developing nations are the main locations for uncontrolled landfills. The unplanned disposal of municipal waste in a low-lying location, typically on the outskirts of the city, is referred to as “open dumping.”
II. *Semi-controlled Landfills* - Daily soil cover and solid waste compaction are done in this procedure to limit odor disturbances or to stop rainwater from seeping into the solid waste being placed.
III. *Engineered Landfills* - Engineered landfills are primarily employed in industrialized nations and are considered the most effective waste management option for disposing of garbage on the earth’s surface soils. Additionally, designed landfills are systems with facilities for controlling landfills, such as liners, systems for collecting and extracting landfill leachate, systems for collecting and extracting landfill gas, and daily and ultimate cover layers (Tchobanoglous et al., 1993).

5. Landfilling Phases

The following are the many phases of degradation that waste products go through after being disposed of as solid trash in landfills (Kjeldsen et al., 2002):

5.1. Initial Aerobic Phase

The oxygen that is present in the waste mass is quickly used up during the first stage of the aerobic phase, and carbon dioxide is created as the temperature of the trash rises. This stage of aerobic conditions lasts for a short while before oxygen appears in the waste. Additionally, after the trash was covered with dirt, oxygen couldn't be added to restart the aerobic phase. Leachates are created during this phase as a result of the compaction of the trash (Tchobanoglous et al., 1993).

5.2. Anaerobic Acid Phase

As the initial phase of the waste matter exhausts the oxygen, an anaerobic phase begins, which leads to fermentation reactions. Landfills create methane and carbon dioxide under anaerobic conditions as cellulose and hemicellulose decompose (Barlaz et al., 1990). The following three types of bacteria are involved in the biodegradation processes:

I. Hydrolytic and fermentative bacteria
This class of bacteria takes part in the hydrolysis of polymeric compounds to produce monosaccharides, which are then fermented to make alcohols and carboxylic acids.

II. Acetogenic bacteria
This group of bacteria transforms the produced carboxylic acids and alcohols into acetate, hydrogen and carbon dioxide.

III. Methanogenic bacteria
Methane and carbon dioxide are produced by methanogenic bacteria, which convert acetate and hydrogen gas (Zehnder, 1982). This process must take place in an environment with a pH close to neutral. The majority of the hydrolytic, fermentative, and acetogenic bacteria are still active during this anaerobic acid phase, which causes carboxylic acids to form and reduce the pH of the leachate. In this acidic phase, leachate's highest levels of chemical oxygen demand (COD) and biochemical oxygen demand (BOD5) are seen (Reinhart and Grosh, 1998). The BOD5/COD ratio is also discovered to be approximately 0.7 (Robinson, 1995). In this acidic environment, several chemicals that are normally insoluble in

leachate are dissolved. There are three steps during the methanogenic phase-

a) Initial Methanogenic Phase

Methanogenic converting the carboxylic acids generated in the anaerobic acid phase into methane and carbon dioxide (Christensen and Kjeldsen, 1989). As the carboxylic acids are used up, the pH of the leachate progressively rises and stays between 6.8 and 8. As the concentration of carboxylic acids falls, so does the concentration of COD, BOD5, and the ratio of BOD5/COD.

b) Stable Methanogenic Phase

The stable methanogenic phase is when the most methane is produced. However, methane generation eventually declines as carboxylic acid concentrations fall. The leachate solution's pH rises and maintains an alkaline state. Leachate contains small levels of COD, primarily in the form of recalcitrant humic and fulvic acids (Christensen et al., 1994). The BOD5/COD ratio continues to fall throughout this phase, reaching as low as 0.1.

c) Aerobic or Humic Phase

No landfills have ever recorded this latter stage of the aerobic condition. The decomposition of the waste will be complete when this aerobic situation supposedly develops. Once the preceding phases pass, air will permeate the waste matter to reach the aerobic phase after the amount of methane generation decreases.

The primary disadvantage of landfilling is the production of extremely contaminated leachates, whose volumetric flow rate, quantity, and chemical makeup are very changeable. Additionally, landfills create risks and negative effects on the environment, which forces authorities to enact ever-stricter regulations.

6. Landfill as Wasteland and Its Reclamation

Landfilling also makes the land unsuitable for any purpose. Land used for landfill cannot be utilized for agriculture, residential, industrial, commercial, recreational or any other legitimate purposes. So, the landfill must also be considered a wasteland and tried for reclamation. Landfill reclamation for municipal solid waste (MSW) is the process of removing previously disposed of materials from a landfill to recover soil, fine materials (referred to above as reclaimed soil), metal, glass, plastic, and other combustibles, as well as the landfill volume itself (Morelli, 1990). In several landfill sites around the US, landfill reclamation has been used sparingly (Murphy & Stessel, 1991; NYSERDA, 1992; Serpa, 2008; US EPA, 1997), and there is growing interested in the method on a global scale (Jones et al., 2013). The main reasons operators have thought about and implemented landfill reclamation at their sites are: (1) To address groundwater contamination issues caused by wastes in old, unlined landfills by removing the source of pollution; (2) to create new capacity for future landfilling activities; and (3) to lower closure costs by decreasing the footprint area of a landfill. The recovery of recyclables, particularly metals, for resale has been noted in some instances as a side benefit of landfill reclamation. Reclamation may be employed in a bioreactor landfill operation to recover stabilized solid waste as part of an integrated approach for sustainable landfilling (Nelson, 1994; Reinhart and Townsend, 1997). Before deciding to carry out the project, it is necessary to assess the reclamation's economic viability. The cost of the project is significantly influenced by several variables, including the composition of the excavated waste (specifically its soil content), the intended use of the recovered materials and reclaimed soil, the quantity and value of the recovered airspace, and the cost of excavating and processing waste. Some information regarding the make-up of the debris from excavation and the caliber of reclaimed soil has been made public (Das et al., 2002; Earle et al., 1999; Flyhammar et al., 1998; Hull et al., 2005; Jain et al., 2005; Kilmer and Tustin, 1999; Zornberg et al., 1999). Scientific Remediation is a process in which contaminants are removed or neutralized scientifically so that they cannot cause further harm to the environment. On-site clean-up is a tried-and-true strategy for reducing environmental concerns from abandoned landfills and eventually reclaiming the area for public use (Sardana, 2022).

Recommendations for remediating Indian landfills for the following purposes, all current and former landfills will be examined, and boreholes will be dug: (i) recovered samples of leachate from the landfill's base; (ii) recovered samples from the soil beneath the landfill's base for permeability and soil characteristics evaluation; and (iii) recovered samples of garbage for waste characterization. At least three boreholes will be sunk, with one borehole drilled for every acre of the landfill area. Leachate samples' quality will be compared to (a) the quality of the groundwater in borewells that are already in place two kilometers from the landfill and (b) the Central Pollution Control Board's (CPCB's) recommendations for the maximum allowable concentrations of contaminants in leachate. The current landfill may continue to operate with biannual leachate quality monitoring in the drilled boreholes if the permeability of the subsurface strata and leachate quality is found to be satisfactory. (f) If it is determined that the leachate does not meet CPCB standards or the local groundwater quality standards, efforts will be made to close the current landfill site and corrective actions will be initiated (Sardana, 2022).

7. Post Closure MSW Rules 2016 in India (Sardana, 2022)

1) The post-closure care of the landfill site must be carried out for at least fifteen years, and the long-term monitoring or care plan must include the following:
   i. Maintaining the integrity and effectiveness of the final cover, making repairs, and preventing run-on and run-off from eroding or otherwise damaging the final cover;
   ii. Monitoring the leachate collection system following the requirement;
   iii. Monitoring of the groundwater in and around the landfill;
   iv. Maintaining and operating the landfill gas collection system to meet the standards.
2) After fifteen years of post-closure monitoring, the use of closed landfill sites may be considered for human settlement or other purposes, but only after confirming that soil stability and gaseous emission and leachate quality assessments meet the required standards.

8. Landfill Remediation Technologies

8.1. Landfill Mining

Landfill mining is an environmentally-friendly technology that combines the concepts of material recycling and sustainable waste management, and it has received a great deal of worldwide attention because of its significant environmental and economic potential in material recycling, energy recovery, land reclamation and pollution prevention (Zhou, et al., 2015). A practical solution for the ecological repair of ancient landfills is material management, which combines municipal waste management and material recycling (Krook and Baas, 2013). Get fertilizer for orchards, landfill mining was first introduced in Israel in 1953, according to Savage et al., (1993). Landfill mining offers the ability to recover minerals and other natural resources from waste materials in comparison to conventional landfill clean-up techniques. After sorting and screening the stored garbage, soil-like materials can be used as organic fertilizer in gardens and green areas, metals and building waste can be recycled, and waste-to-energy technology (e.g., incineration, gasification, pyrolysis, plasma technologies, combinations) can be used to treat materials with a high calorific value for energetic valorization and material recovery (Krook et al., 2012; Bosmans et al., 2013; Jones et al., 2013). Landfill mining is considered if environmental pollution is urgent, land is scarce, or combustible waste treatment and energy recovery facilities are available (Zhao et al., 2007; Jain et al., 2013; Jones et al., 2013). Recent research mostly focused on the components and characteristics of stored waste and waste-to-energy/material (WtE) technologies (Hull et al., 2005; Joseph et al., 2007; Quaghebeur et al., 2013; Bosmans et al., 2013). Assessing the economic feasibility is an important issue when making decisions concerning a landfill mining project, but only a few studies have focused on the economic issue of landfill mining. The US-EPA (1997) developed a framework for assessing the potential economic benefits and the capital and operational costs of landfill mining projects. Decision-makers can choose landfill mining projects with the best market potential by using a step-by-step methodology developed by Van der Zee et al., (2004) that combines cost-benefit analysis (CBA) and multi-criteria analysis (MCA). This methodology was then used by Van Passel et al., (2013) to construct a CBA simulation tool that took into account both the advantages of waste to material (WtM), waste to energy (WtE), and land reclamation as well as the costs of landfill mining, such as excavation, sorting and pre-treatment, incineration, and contingency. Ayalon et al., (2006) and Marella and Raga (2014) employed the contingent valuation technique (CVM) to evaluate the social advantages of landfill mining and rehabilitation. Landfill mining is done in three phases (CSE, 2020):

I. Stabilization
The dumping site is changed periodically, divided into windrows of similar size, added with bio-culture, and given a deodorizer spray. This stage eliminates odor, lessens moisture and flies, kills pathogens, accelerates the breakdown of waste into carbon dioxide and water vapor, and generates biological heat inside the trash, which helps to dry it out and reduces its volume by 35–40%. The waste is sufficiently dried by this method for screening. When there is no longer any production of heat, landfill gas, or leachate, the trash is deemed stable. To check that the garbage has been stabilized properly, samples are taken from the heap and subjected to a germination test.

II. Sorting and segregation
In this stage, stabilized excavated landfill debris is divided into soil, rocks, and combustibles. The aggregates and other heavy construction trash and debris are separated using a series of trommels and labor.

III. Disposal
The non-combustible fractions, such as dirt and stones, are disposed of in this phase to be processed into finer aggregates or used as earth fill. Refuse-derived fuel (RDF), which is what the combustible fractions are known as is typically delivered for co-processing in cement plants.

8.2. Landfill Capping and/or Phytocap

Using this technique, the dumpsite is first leveled, covered with soil, and equipped with leachate management, gas collection, and surface drainage systems. These actions transform the garbage site into a green area with environmental monitoring devices. This is utilized when there are no other effective reclamation options and bioremediation would be prohibitively expensive, there would be excessive levels of pollution, or the material that would emerge out of the legacy dump site would be unpredictable. Three layers are used to cap a landfill: an upper vegetative (topsoil) layer, a drainage layer, and a layer with low permeability that is made of synthetic material and sits on top of two feet of compacted clay. Although the functioning of the cap is dependent on the site's climatic circumstances, capping has a lifetime of 50 to 100 years. Changes in precipitation and air temperature, as well as if the area is vulnerable to subsidence and earthquakes, can cause caps to fracture and disintegrate. The top needs to be thick enough to support animal burrows and vegetative roots to avoid freezing.

8.3. Landfills Surcharging

Surcharging is the process of adding weight to the landfill. Typically, stocks of dirt are used for this (clean soil so that it can be reused). Settlements become apparent the heavier and longer a stockpile is in place. The operation for recycling green and wood trash on top of the dump can also help the settlement. Surcharging is often done once a landfill reaches its highest elevation.

8.4. Soil Vapor Extraction and Air Sparging

Both soil vapor extraction (SVE) and air sparging are in-situ remediation techniques that are used to remove vapors from polluted soil and plumes, respectively. SVE and air sparging are two processes that are utilized simultaneously on-site. Solvents, fuels, and volatile organic substances are easily removed in this way. The technique entails building two different kinds of wells—air injection wells and extraction wells—around the landfill. While an air injection well pumps air into the ground, the extraction well creates a vacuum to bring the vapors to the surface. The air injection encourages aerobic bacterial growth, which aids in microbial breakdown. If the injected air is heated, the chemicals evaporate more quickly. The amount and depth of the pollution, the type of soil, and the chemical concentrations in the soil and groundwater all affect how quickly full remediation may be accomplished. SVE and air sparging are both safe, but full remediation may take years. Contrarily, these solutions are quicker than relying only on natural processes.

8.5. Co-Treatment

8.5.1. Landfill Leachate with Sewage in a Wastewater Treatment Plant

Leachate from this is often sent to a wastewater treatment facility. Leachate management solutions include pre-treatment before release to a wastewater treatment facility or treatment on-site followed by discharge to a neighboring stream. The most frequent strategy is to join a nearby sewer line. The majority of municipal wastewater treatment facilities use aerobic biological treatment, which was developed primarily to remove suspended solids and biodegradable organic matter from sewage (for example, the activated sludge process). As a result, emerging pollutants and refractory organics may not be sufficiently eliminated from leachate. It's vital to keep in mind that although a lot of large sewage can dilute the persistent pollutants in leachate, these pollutants aren't eliminated. Additionally, dangerous substances in leachate (such as ammonia and heavy metals) may interfere with microbial activity and result in abnormal operation of a wastewater treatment facility. Additionally, sewer lines may not be able to connect to local treatment facilities due to accessibility issues, capacity issues, or other issues.

8.5.2. Bioreactor Landfills

By cycling leachate back into filled cells, the microbial activity in a bioreactor landfill is intentionally boosted. Leachate recirculation lessens the amount of leachate needed for treatment while supplying moisture and/or oxygen to promote the microbial breakdown of solid wastes. A bioreactor landfill can be either aerobic or anaerobic to recover landfill space. While aerobic bioreactor landfills pump both air and leachate into the waste, anaerobic bioreactor landfills just inject leachate. The aerobic bioreactor accelerates microbial digestion rates, leading to faster waste settlement when compared to anaerobic bioreactor landfills. On the other hand, anaerobic bioreactor landfills generate more methane gas. As a result, it's a fantastic choice for projects involving energy recovery.

8.6. *In Situ* Aeration

Municipal Solid Waste (MSW) landfill in situ aeration has gained popularity, and several projects have been completed in the past ten years globally. Thorough analyses on the utilization of in situ aeration of MSW dumps are available (Rich et al., 2008; Ritzkowski and Stegmann, 2012), offering an overview of preexisting ideas and identifying several instances in which various objectives were pursued and various approaches were employed. The reported cases can be divided into two categories based on the reasons for using landfill aeration:

- quickening the biological stabilization of trash to cut down on post-closure care at landfills and lower both current emissions and probable residual emissions;
- pre-treating the trash before mining the dump.

For the evaluation of the effects of aeration on waste biological stability, leachate, and biogas emissions, numerous lab- and full-scale applications of in situ aeration were conducted (Cossu et al., 2001; Heyer et al., 2005; Prantl et al., 2006; Raga and Cossu, 2013; Ritzkowski et al., 2006). Leachate recirculation and intermittent aeration's effects on leachate quality were studied in a full-scale application by Oncü et al., (2012); the key issue of whether lab-scale studies are appropriate to represent what occurs in full-scale applications has recently been addressed (Hrad et al., 2013). Ritzkowski and Stegmann (2012) propose a methodological approach based on carbon balance including the assessment of the biodegradable organic carbon (BOC), the carbon discharge via leachate and off gas, and the amount of bio-convertible carbon. They discuss important issues related to the evaluation of the time needed for landfill bio-stabilization employing in situ aeration and of the related target values for the relevant waste stability indexes. It has become clear that in situ aeration is essential for preventing the smells and dangers associated with methane during waste mining. If the appropriate steps are taken to restrict biogas migration in the excavation region, the target can be met even for greater values of the stability indexes, which would result in a shorter aeration time (Cossu et al., 2003). However, if future disposal in new landfill sectors is anticipated, extended aeration may also lead to the mining of a stable organic fraction that doesn't require additional treatment for biological stabilization following excavation. One of the suggested actions for the good management of transient storage facilities for certain waste streams is in situ aeration (Wagner and Bilitewski, 2009).

A company promoted and involved by the University of Padua, developed the Airflow patented technology, which is based on low-pressure air injection and gas extraction, with simultaneous pneumatic extraction of landfill leachate, to apply in situ aeration to MSW landfills in Italy (Cossu et al., 2003a, 2005). An Italian MSW landfill (Landfill C) was the subject of a reclamation project that included preliminary landfill conditioning using the Airflow technology, landfill mining with material and energy recovery, construction of new landfill sectors providing for disposal capacity for two decades, and eventual landscaping for the creation of new functionalities through the construction of a public park in the landfill area. (Cossu, 2003). The data obtained imply that the conditioning before excavation of the remaining landfill sectors should be structured differently: the aeration wells should be employed for an initial enhanced leachate and biogas extraction, to make the subsequent aeration step more efficient. Despite the low permeability, the option of recirculating leachate in specific areas to prevent drying effects and improve leachate biodegradation processes should be taken into consideration. This may be done in conjunction with brief high-pressure air pulses, as described in Oncü et al., (2012), to improve air distribution and shift leachate pockets.

The evaluation of the contamination is the initial step in the landfill rehabilitation process. Often, the procedure starts with an environmental site assessment. The way the site is used and the items deposited there will determine the evaluation method, kind of sample, and chemical analysis to be carried out. Even though the current land use seems to be safe, neighboring sites that are owned by the same company or close-by sites that have been reclaimed, leveled, or filled frequently contain contaminants. It's also important to deal with off-site pollution, which frequently results from decades' worth of emissions to the air, groundwater, and soil. The topsoil, groundwater, surface water, and ceiling dust of neighboring homes should all be assessed both before and after any repair. The fourth requirement is that transportation and remediation expenses as well as the environmental impact and social acceptance be taken into account (Raga, 2014).

9. Leachate and Gas

The natural consequence of the breakdown of organic waste in landfills is landfill gas (LFG). Methane, the main component of natural gas, makes up around 50% of LFG, along with 50% carbon dioxide (CO_2) and a little amount of organic molecules that aren't methane.

9.1. Collecting and Treating Landfill Gas

Landfill gas can be captured, processed, and used as a renewable energy source as an alternative to escaping into the atmosphere. Landfill Gas is used to stop methane from evaporating into the atmosphere and causing local smog and global climate change, as well as to eliminate odors and other dangers related to Landfill Gas emissions. Energy projects using landfill gas also generate income and job opportunities in the local area and beyond.

9.2. Closure of Remediation Landfill

i. Compacting of land for Waste
ii. Before the waste processing, the waste shall be sent to the sanitary landfill.
iii. Before monsoon season, an intermediate cover of 40-65 cm thickness of soil shall be placed on the landfill with proper compaction and grading
iv. The final cover shall be designed to minimize infiltration and erosion after the completion of the landfill.
v. The final cover shall have a barrier soil layer comprising 60 cm of clay with a proper drainage layer of 15cm on top of its vegetative layer of 45 cm to support natural plant growth and minimize erosion.

The landfill site must be maintained after it has been closed for at least fifteen years, and a long-term monitoring or maintenance plan must include the following:

i. Monitoring the leachate collection system following the requirement;
ii. Monitoring the groundwater in and around the landfill;
iii. Maintaining and operating the landfill gas collection system to meet the standards.
iv. Maintaining the integrity and effectiveness of the final cover, making repairs, and preventing run-on and run-off from eroding or otherwise damaging the final cover.

After fifteen years of post-closure monitoring, the use of closed landfill sites can be considered for human settlement or other purposes, but only after confirming that soil stability and gaseous emission and leachate quality assessments meet the required standards.

9.3. Economies of Managing Dumpsites

According to CPCB the cost of on-site bioremediation and biomining is Rs 400–700 per cubic meter and the average cost of processing one tonne of legacy waste is Rs 750–900, depending on the age of the dumpsite irrespective of capital cost (CPCB, 2019).

10. Benefits and Drawbacks of Landfill Reclamation

10.1. Potential Benefits

i. By extracting salvageable elements and reducing trash volume through combustion and compaction, landfill reclamation prolongs the life of the existing facility.
ii. If there are markets for recycled materials like ferrous metals, aluminum, plastic, and glass, these commodities can be sold.
iii. Reclaimed soil can be used locally as a daily cover material on other landfill cells, saving the expense of importing cover soil and lowering operational costs or earning cash from its sale. Reclaimed dirt utilized in other uses, such as construction fill, may also have a market.
iv. Combustible recycled garbage can be combined with fresh waste and burned to create electricity at MWCs.
v. The facility operator may be able to minimize the cost of closing the landfill or free up the area for other purposes by shrinking the landfill's "footprint" through cell reclamation.
vi. Older landfills can have liners and leachate collection systems installed. If these mechanisms are already in place, they can be examined and fixed. Additionally, it is possible to remove and manage hazardous trash more securely.

10.2. Potential Drawbacks

i. Particularly at older landfills, hazardous materials that may be discovered during reclamation activities are subject to strict management and disposal standards. Hazardous waste management expenses can be rather significant, but they may lower long-term responsibility.
ii. The emission of gases during cell excavation creates several possible issues. Explosions and flames can be started by the methane and other gases released by decomposing waste. When inhaled in large quantities, the extremely combustible and pungent gas hydrogen sulfide can be lethal.
iii. When a landfill region is dug, surrounding cells may lose their structural integrity and sink or collapse into the excavated area.
iv. Excavators and loaders, which are used in reclamation processes, have a shorter usable life due to the high density of the trash that needs to be handled. Also, MWC equipment (such as grates and air pollution control systems) may wear down more quickly due to the high particle concentration and abrasive character of recovered waste.

There are some examples of cities where biomining and bioremediation of legacy waste are going on successfully on landfill sites like Bhalswa landfill site (Delhi), Hyderabad, Kumbakonam, Vijayawada, Ambikapur, Indore etc. in India. (CSE, 2020).

Conclusion

Wasteland is the land that is not under any use due to different reasons and may be of culturable and Unculturable types and covers a large part of a country. Anthropogenic activities produce several undesirable, unwanted materials known as solid waste. Waste management practices include thermal or biological methods and practices like reuse, recycling, waste to energy conversion etc. The landfill is the disposal site for municipal solid waste. The landfill may be of area, trench, or depression type. The landfill also acts as a wasteland because it does not have any use. Landfill results in several environmental and ecological problems like generation of gas and leachate production, subsidence, fire hazard, groundwater contamination, breeding ground of infectious diseases etc. to cope with these problems it is necessary to remediate and reclaim the landfill sites to transform the wasteland into useful land. Common methods available for landfill reclamation include landfill mining, phytocapping, in situ aeration and Landfills Surcharging etc. Several legacy landfills around the world have been successfully reclaimed through these processes and many are under operation for the reclamation.

References

Agarwal, U. P. (1986). Wastelands and their reclamation in Himachal Pradesh. In *Proceedings of the Workshop on Wasteland Development in Himachal Pradesh*, April 17–18 p. 32. Society of Wasteland Development.

Annepu, R. K. (2012). *Sustainable solid waste management in India* [M.Sc. Thesis]. Earth Engineering Center, Columbia University.

Ayalon, O., Becker, N., & Shani, E. (2006). Economic aspects of the rehabilitation of Hiriya landfill. *Waste Management*, *26(11)*, 1313-1323.

Barlaz, M. A., Ham, R. K., Schaefer, D. M., & Isaacson, R. (1990). Methane production from municipal refuse: A review of Enhancement Techniques and Microbial Dynamics. *Critical Reviews in Environmental Control*, *19*(6), 557–584. https://doi.org/10.1080/10643389009388384.

Bhumla, D. R., & Khare, A. (1984). Estimates of wastelands in India. In *Wasteland diagnosis and treatment*(ed.),Yadav, H. R. p. 28. Concept Publishing Company.

Bosmans, A., Vanderreydt, I., Geysen, D., Helsen, L. (2013). "The crucial role of waste to energy technologies in enhanced landfill mining: a technology review." *J. Clean. Product* 55:1023.

Christensen, T. H., & Kjeldsen, P. (1989). Basic biochemical processes in landfills. In T. H. Christensen, R. Cossu & R. Stegmann (Eds.), *Sanitary landfilling: Process, technology and environmental impact* p. 29. Academic Press.

Christensen, T. H., Kjeldsen, P., Albrechtsen, H. J., Heron, G., Nielsen, P. H.,Bjerg, P. L., & Holm, P. E. (1994). Attenuation of landfill leachate pollutants in aquifers. *Critical Reviews in Environmental Science and Technology*, *24*(2), 119–202. https://doi.org/10.1080/10643389409388463.

Cossu, R. (2003). Environmental reclamation of the Campodarsego (PD) controlled waste landfill. In *Proceedings of the Sardinia 2003, Ninth International Waste Management and Landfill Symposium*. Cagliari, Italy, October 6–10.

Cossu, R., Lavagnolo, M. C., & Raga, R. (2001). In situ stabilisation of old landfills: Lab scale and field tests. In *Proceedings of the Sardinia 2001, Eighth International Waste Management and Landfill Symposium, Cagliari, Italy*, October 1–5, Vol IV, 531-540.

Cossu, R., Pivato, A., & Raga, R. (2003b). Preliminary risk assessment of old landfills in Italy. *WIT Transactions on Biomedicine and Health*, *8*, 195–203. https://doi.org/10.2495/EHR030201.

Cossu, R., Raga, R., & Rossetti, D. (2003a). Full scale application of in situ Aerobic stabilization of old landfills. In *Proceedings of the Sardinia 2003, Ninth International Waste Management and Landfill Symposium, Cagliari, Italy*, October 6–10.

Cossu, R., Sterzi, G., & Rossetti, D. (2005). Full-scale application of aerobic in situ stabilization of an old landfill in north Italy. In *Proceedings of the Sardinia 2005, Tenth International Waste Management and Landfill Symposium, Cagliari, Italy*, October 3–7. 237-238.

CPCB. (2019). *Operation and maintenance cost for bio-remediation and bio-mining. Guidelines for disposal of legacy waste (old municipal solid waste)* p. 25. Ministry of the Environment and Climate Change, Government of India.

CPHEEO. (2000). *Manual on municipal solid waste management*. Central Public Health and Environmental Engineering Organization. Ministry of Urban Development. Govt. of India. New Delhi.

CSE. (2020). "Clean It Right Dumpsite Management in India." Centre for Science and Environment. *Usha Murti Enterprises*, New Delhi. 17-18.

Das, K. C., Smith, M. C., Gattie, D. K., & Hale Boothe, D. D. H. (2002). Stability and quality of municipal solid waste compost from a landfill aerobic bioreduction process. *Advances in Environmental Research*, *6*(4), 401–409. https://doi.org/10.1016/S1093-0191(01)00066-1.

Dutta, A. (2015). Increasing the sustainable energy recovery from organic fraction of MSW in India's urban cities. *Organized by departmental of mechanical engineering*. Paper presented in India-UK Scientific Seminar on Sustainable energy recovery from waste biomass (SERB), February 12–14. Jadavpur University and Kingston University.

Earle, C. D. A., Rhue, R. D., & Earle, J. F. K. (1999). Mercury in a municipal solid waste landfill. *Waste Management and Research*, *17*, 305–312.

Flyhammar, P., Tamaddon, F., & Bengtsson, L. (1998). Heavy metals in a municipal solid waste deposition cell. *Waste Management and Research: The Journal for a Sustainable Circular Economy*, *16*(5), 403–410. https://doi.org/10.1177/0734242X9801600502.

Gautam, S. P., Bundela, P. S., Pandey, A. K., Awasthi, M. K., & Sarsaiya, S. (2010). Composting of municipal solid waste of Jabalpur City. *Global Journal of Environmental Research*, *4*(1), 43–46.

Gilbert, P., Ryu, C., Sharifi, V., & Swithenbank, J. (2009). Effect of process parameters on pelletisation of herbaceous crops. *Fuel*, *88*(8), 1491–1497. https://doi.org/10.1016/j.fuel.2009.03.015.

Gupta, N., Yadav, K. K., & Kumar, V. (2015). A review on current status of municipal solid waste management in India. *Journal of Environmental Sciences*, *37*, 206–217. https://doi.org/10.1016/j.jes.2015.01.034.

Gupta, V. K., Sadegh, H., Yari, M., Shahryari Ghoshekandi, R., Maazinejad, B., & Chahardori, M. (2015). Removal of ammonium ions from wastewater. A short review in development of efficient methods. Glob. *Journal of Environmental Science and Management*, *1*(2), 149–158.

Heyer, K. U., Hupe, K., Ritzkowski, M., & Stegmann, R. (2005). Pollutant release and pollutant reduction – Impact of the aeration of landfills. *Waste Management*, *25*(4), 353–359. https://doi.org/10.1016/j.wasman.2005.02.007.

Holm-Nielsen, J. B., Al Seadi, T., & Oleskowicz-Popiel, P. (2009). The future of anaerobic digestion and biogas utilization. *Bioresource Technology*, *100*(22), 5478–5484. https://doi.org/10.1016/j.biortech.2008.12.046.

Hrad, M., Gamperling, O., & Huber-Humer, M. (2013). Comparison between lab- and full-scale applications of in situ aeration of an old landfill and assessment of long-term emission development after completion. *Waste Management*, *33*(10), 2061–2073. https://doi.org/10.1016/j.wasman.2013.01.027.

Hull, R. M., Krogmann, U., & Strom, P. F. (2005). Composition and characteristics of excavated materials from a New Jersey landfill. *Journal of Environmental Engineering*, *131*(3), 478–490. https://doi.org/10.1061/(ASCE)0733-9372(2005)131:3(478).

Jain, P., Kim, H., & Townsend, T. G. (2005). Heavy metal content in soil reclaimed from a municipal solid waste landfill. *Waste Management*, *25*(1), 25–35. https://doi.org/10.1016/j.wasman.2004.08.009.

Jain, P., Townsend, T. G., & Johnson, P. (2013). Case study of landfill reclamation at a Florida landfill site. *Waste Management*, *33*(1), 109–116. https://doi.org/10.1016/j.wasman.2012.09.011.

Jones, P. T., Geysen, D., Tielemans, Y., Van Passel, S., Pontikes, Y., Blanpain, B., Quaghebeur, M., & Hoekstra, N. (2013). Enhanced landfill mining in view of multiple resource recovery: A critical review. *Journal of Cleaner Production*, *55*, 45–55. https://doi.org/10.1016/j.jclepro.2012.05.021.

Joseph, K., Esakku, S., & Nagendran, R. (2007). Mining of compost from dumpsites and bioreactor landfills. *International Journal of Environmental Technology and Management*, *7*(3/4), 317–325. https://doi.org/10.1504/IJETM.2007.015149.

Karmakar, S., De, S., Maiti, S. K., Baidya, R., Manna, A. K., Hazra, T., Dutta, A., & Ghsoh, S. K. (2014). Review of the status of management of MSW focusing MSW rules 2000 and document published as draft MSW rule 2013 in India. In *Proceedings of the 4th Icon SWM* 2014 on Waste Management & Resource Utilisation organized by Centre for Quality Management System, Jadavpur University. *Commissioner and director of Municipal Administration, Govt. of Andhra Pradesh and International Society of waste management, air and water at Hyderabad*, January 28–30. Oxford Publishing House, Kolkata. 137-142.

Kilmer, K. S., & Tustin, J. (1999). Rapid landfill stabilization and improvements in leachate quality by leachate recirculation. In *Proceedings of the SWANA's 4th Annual Landfill Symposium*, June 28–30, Denver. CO.

Kjeldsen, P., Barlaz, M. A., Rooker, A. P., Baun, A., Ledin, A., & Christensen, T. H. (2002). Present and long-term composition of MSW landfill leachate: A review. *Critical Reviews in Environmental Science and Technology*, *32*(4), 297–336. https://doi.org/10.1080/10643380290813462.

Krook, J., & Baas, L. (2013). Getting serious about mining the technosphere: A review of recent landfill mining and urban mining research. *Journal of Cleaner Production*, *55*, 1–9. https://doi.org/10.1016/j.jclepro.2013.04.043.

Krook, J., Svensson, N., & Eklund, M. (2012). Landfill mining: A critical review of two decades of research. *Waste Management*, *32*(3), 513–520. https://doi.org/10.1016/j.wasman.2011.10.015.

Late, A., & Mule, M. B. (2013). Composition and characterization study of solid waste from Aurangabad city. *Universal Journal of Environmental Research and Technology*, *3*(1), 55–60.

Marella, G., & Raga, R. (2014). Use of the contingent valuation method in the assessment of a landfill mining project. *Waste Management*, *34*(7), 1199–1205. https://doi.org/10.1016/j.wasman.2014.03.018.

McKendry, P. (2002). Energy production from biomass (part 3): Gasification technologies. *Bioresource Technology*, *83*(1), 55–63. https://doi.org/10.1016/S0960-8524(01)00120-1.

Morelli, J. (1990). Landfill reuse strategies. *BioCycle* (April), *31*(4), 60–61.

Murphy, R. J., & Stessel, R. I. (1991). *Optimization of landfill mining: Interim final report- Phase I*. Center for Solid and Hazardous Waste Management.

Narayana, T. (2009). Municipal solid waste management in India: From waste disposal to recovery of resources? *Waste Management*, *29*(3), 1163–1166. https://doi.org/10.1016/j.wasman.2008.06.038.

Ndegwa, P. M., & Thompson, S. A. (2001). Integrating composting and vermicomposting in the treatment and bioconversion of biosolids. *Bioresource Technology*, *76*(2), 107–112. https://doi.org/10.1016/S0960-8524(00)00104-8.

Nelson, H. (1994). Landfill reclamation strategies. *BioCycle* (October), *35*(10), 41–44.

New York State Energy Research and Development Authority. (1992). *Town of Edinburg landfill reclamation demonstration project* Final Report Prepared for the New York State Research and Development Authority.

NRSA. (1991). Guidelines to use wasteland maps. *National Remote Sensing Agency, Hyderabad*, 1-*72*.

NWDB. (1986). 'Wasteland: Definition and Database.' report of the technical task group. *National Wasteland Development Board, Government of India, New Delhi*, 1-*56*.

Öncü, G., Reiser, M., & Kranert, M. (2012). Aerobic in situ stabilization of Landfill Konstanz Dorfweiher: Leachate quality after 1 year of operation. *Waste Management*, *32*(12), 2374–2384. https://doi.org/10.1016/j.wasman.2012.07.005.

Prantl, R., Tesar, M., Huber-Humer, M., & Lechner, P. (2006). Changes in carbon and nitrogen pool during in-situ aeration of old landfills under varying conditions. *Waste Management*, *26*(4), 373–380. https://doi.org/10.1016/j.wasman.2005.11.010.

Quaghebeur, M., Laenen, B., Geysen, D., Nielsen, P., Pontikes, Y., Van Gerven, T., & Spooren, J. (2013). Characterization of landfilled materials: Screening of the enhanced landfill mining potential. *Journal of Cleaner Production*, *55*, 72–83. https://doi.org/10.1016/j.jclepro.2012.06.012.

Raga, R., & Cossu, R. (2013). Bioreactor tests preliminary to landfill in situ aeration: A case study. *Waste Management*, *33*(4), 871–880. https://doi.org/10.1016/j.wasman.2012.11.014.

Raga, R., & Cossu, R. (2014). Landfill aeration in the framework of a reclamation project in Northern Italy. *Waste Management, 34*(3), 683–691. https://doi.org/10.1016/j.wasman.2013.12.011.

Rajput, R., Prasad, G., & Chopra, A. K. (2009). 'Scenario of solid waste management in present Indian context.' Casp. *Journal of Environmental Sciences*, *7*(1), 45–53.

Ranade, P. S. (2009). *Infrastructure development and its environmental impact- study of Konkan railway*. Concept Publication Company 59.

Reinhart, D. R., & Grosh, C. J. (1998). *Analysis of Florida MSW landfill leachate quality* Report 97-3. Florida Center for Solid and Hazardous Waste Management.

Reinhart, D. R., & Townsend, T. G. (1997). *Landfill bioreactor design and operation.* Lewis Publishers.

Rich, C., Gronow, J., & Voulvoulis, N. (2008). The potential for aeration of MSW landfills to accelerate completion. *Waste Management, 28*(6), 1039–1048. https://doi.org/10.1016/j.wasman.2007.03.022.

Ritzkowski, M., & Stegmann, R. (2012). Landfill aeration worldwide: Concepts, indications and findings. *Waste Management, 32*(7), 1411–1419. https://doi.org/10.1016/j.wasman.2012.02.020.

Ritzkowski, M., Heyer, K. U., & Stegmann, R. (2006). Fundamental processes and implications during in situ aeration of old landfills. *Waste Management, 26*(4), 356–372. https://doi.org/10.1016/j.wasman.2005.11.009.

Robinson, H. D. (1995). *'The technical aspects of controlled waste management. A review of the composition of leachates from domestic wastes in landfill sites.' Report for the UK Department of the Environment*. Waste Science and Research, Aspinwall, and Company, Ltd.

Sardana, N. (2022). *'Scientific Landfill Remediation.' Training Manual, Climate Smart Cities Assessment Framework Waste Management, Climate Centre for Cities, NIUA in association with UNEP under the Counter Measures (II).*

Savage, G. M., Golueke, C. G., & Von Stein, E. L. (1993). Landfill mining: Past and present. *BioCycle*, *34*, 58–61.

Serpa, L. (2008). Peer-reviewed feature Clovis landfill reclamation project. *MSW Management* (May June).

Shafi, M. (1968). *The problems of waste-lands in India*, *XV*. International Geographical Congress. Aligarh Muslim University Geographical Society, India, 64-73.

Sinha, B. P., & Dubey, B. P. (1986). Status, distribution and development of wastelands in Bihar, dimensions of wastelands development. In *Proceedings of the National Seminar on Wastelands Development, New Delhi,* 23–30.

Stamp, L. D. (1948). *The land of Britain: Its use and misuse*. John Willey and Sons Ltd, London, 433.

Tchobanoglous, G., Theisen, H., & Vigil, S. (1993). *Integrated Solid-Waste Management: Engineering principles and management issues* p. 978. McGraw-Hill, Inc.

US EPA. (1997). *Landfill reclamation* EPA530-F-97-001, *5305w*. Office of Solid Waste and Emergency Response.

Van der Zee, D. J., Achterkamp, M. C., & de Visser, B. J. (2004). Assessing the market opportunities of landfill mining. *Waste Management*, *24*(8), 795–804. https://doi.org/10.1016/j.wasman.2004.05.004.

Van Passel, S., Dubois, M., Eyckmans, J., de Gheldere, S., Ang, F., Tom Jones, P., & Van Acker, K. (2013). The economics of enhanced landfill mining: Private and societal performance drivers. *Journal of Cleaner Production*, *55*(1), 92–102. https://doi.org/10.1016/j.jclepro.2012.03.024.

Wagner, J., & Bilitewski, B. (2009). The temporary storage of municipal solid waste – Recommendations for a safe operation of interim storage facilities. *Waste Management*, *29*(5), 1693–1701. https://doi.org/10.1016/j.wasman.2008.11.018.

Wastelands atlas of India, (2019). *Change analysis based on temporal satellite data of.* (2008–9 and 2015–16). Department of land resource, ministry of rural development. Government of India.

Yadav, I. C., & Devi, N. L. (2009). 'Studies on municipal solid waste management in Mysore city -a case study.' *Report and Opinion*, *1*(3), 15–21.

Yang, H., Yan, R., Chen, H., Lee, D. H., & Zheng, C. (2007). Characteristics of hemicellulose, cellulose and lignin pyrolysis. *Fuel*, *86*(12–13), 1781–1788. https://doi.org/10.1016/j.fuel.2006.12.013.

Zaman, A. U. (2009). Life cycle environmental assessment of Municipal solid waste incinerator municipal solid waste to energy technologies. *Global Journal of Environmental Research*, *3*, 155–163.

Zehnder, A. J. B., Ingvorsen, K., & Marti, T. (1982). Microbiology of methane bacteria. In *Anerrobic digestion, 1981. Proceedings of the Second International Symposium on Anaerobic Digestion, Held in Travemünde, Federal Republic of Germany. On* 6-11 September, International Symposium on Anaerobic Digestion and Hughes, D. E., Eds.,Amsterdam, The Netherlands. Elsevier Biomedical Press B.V. 45-68.

Zhao, Y., Song, L., Huang, R., Song, L., & Li, X. (2007). Recycling of aged refuse from a closed landfill. *Waste Management and Research*, *25*(2), 130–138. https://doi.org/10.1177/0734242X070740.

Zhou, C., Gong, Z., Hu, J., Cao, A., & Liang, H. (2015). A cost–benefit analysis of landfill mining and material recycling in China. *Waste Management*, *35*, 191–198. https://doi.org/10.1016/j.wasman.2014.09.029.

Zornberg, J. G., Jernigan, B. L., Sanglerat, T. R., & Cooley, B. H. (1999). Retention of free liquids in landfills undergoing vertical expansion. *Journal of Geotechnical and Geoenvironmental Engineering*, *125*(7), 583–594. https://doi.org/10.1061/(ASCE)1090-0241(1999)125:7(583).

Chapter 12

The Working of Computer Simulation Models in Deciphering Environmental Problems

Rajesh Kumar Rai *

Department of Agricultural Engineering, Udai Pratap College (Autonomous Institution), Varanasi, India

Abstract

The environment encompasses highly intricate natural phenomena, which become even more challenging to comprehend when spatial and temporal variables are incorporated into the conventional parameters of the study. Consequently, computer simulation models (CSMs) are in high demand within the scientific community. Typically, environmental models address climatic factors and the interconnectedness of physical, chemical, and biological elements in nature. This chapter aims to provide a concise yet informative overview of the functioning of environmental CSMs. It explores how physical data is processed within the framework of a geographical information system (GIS), how data is selected and processed in CSMs, and how scientific conclusions are drawn using these environmental models. These are the types of questions that arise in the minds of researchers when embarking on their initial research projects in this field. Additionally, these models are utilized in studies concerning climate change to examine current trends and predict future impacts in the study area. While this chapter does not offer a comprehensive or exhaustive discussion on CSMs, it serves as a stepping stone toward understanding the models commonly employed in the scientific community.

---

* Corresponding Author's Email: rkrai_agre@yahoo.com.

In: Conserving, Halting and Reinstating Degraded Natural Resources …
Editors: Rajani Srivastava and Pankaj Kumar
ISBN: 979-8-89113-067-8

Keywords: climate change study; computer simulation; environmental models; Geographical Information System (GIS); Graphical User Interface (GUI)

1. Introduction

In the early stages of scientific development, researchers focused on developing derived equations, which eventually led to the creation of empirical equations. These empirical equations served as practical tools for describing natural processes in mathematical terms. With the growing popularity of computer usage in research, problems were translated into mathematical representations and corresponding computer programs were written using appropriate source codes. These extensive computer programs came to be known as computer models or simply models. Generally, models are computer-based programs specifically designed to address specialized scientific problems within a broad area of interest. These models rely on mathematical or empirical equations that serve as abstractions of real-world phenomena. Figure 1 presents a generic diagram of a computer model.

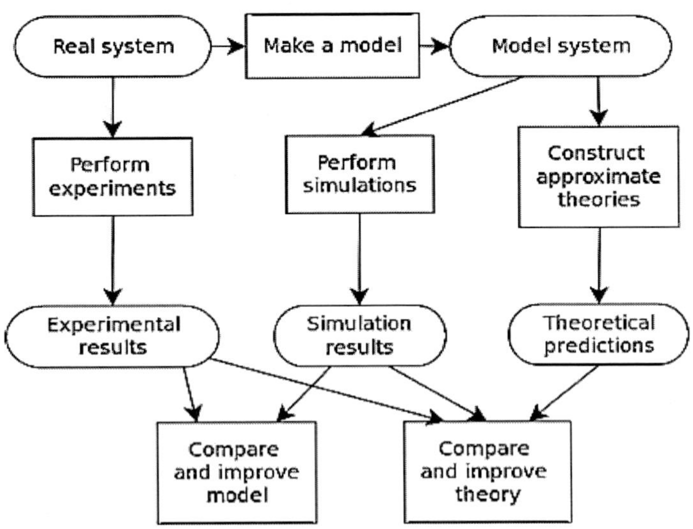

Figure 1. Process of building a computer model (adopted from Allen and Tildesley, 1987).

Environmental models process climate data including rainfall, air temperature, solar radiation, wind velocity (both speed and direction), and humidity (mainly relative humidity). These models incorporate real-world spatial and temporal data to analyze climatic conditions over a specific region and time period (typically 30 years). Dynamic models, which utilize differential equations, are commonly employed to describe the physical environment and its transformations. Examples of dynamic models include the surface flow of rivers (discharge), 2D or 3D groundwater flow equations, and flood forecasting. Hydrological models are a specific category of models used to simulate the land phase of the hydrological cycle. Given the strong connection between agriculture and the environment, agro-hydrological simulation is currently popular. It is important to note that ecosystem modeling encompasses a broader range of studies, such as forest ecosystems, ocean ecosystems, and desert ecosystems, as it considers the interaction between biotic and abiotic factors.

1.1. Working Principle of Major Computer Models

Generally, computer models are lengthy programs designed to solve mathematical problems that require input data. This input data typically consists of climatic data, spatial and temporal data, as well as numerous parameters and constants associated with mathematical equations. Hydrological models, specifically, are computer programs that are predominantly based on conceptual and physical processes. These models are comprehensive and data-driven, operating within a hydrological basin or watershed. When the model program is executed using files specifically created to provide input data, it produces predefined results as output files. In simple terms, this can be understood as follows:

$$\text{INPUT} \Rightarrow \text{MODEL} \Rightarrow \text{OUTPUT}$$

These programs incorporate various experimental constants and assumed or experimental parameters, leading to the inherent uncertainty in their predictions. Therefore, it is necessary to adjust the model parameters and calibrate them to find a unique set of values that accurately describe the problem at hand. Additionally, the chosen statistical model needs to be validated to determine its appropriateness.

To demonstrate the effectiveness of the models and assess the level of satisfaction with the simulations, techniques such as parameter identification and estimation, model verification, sensitivity analysis, and uncertainty analysis are employed (Muleta & Nicklow, 2005).

While these models aim to address a wide range of scientific problems, they often fall short of covering the entire domain of their respective fields. Consequently, researchers commonly combine multiple models to overcome the limitations of each model. To facilitate this, the models need to be inter-compatible or interoperable, allowing easy coupling or integration with one another (referred to as model coupling or integration).

2. Programming Languages Used in Source Code

The programming language plays a crucial role for developers. To build precise climate models, developers require a dependable, resilient, and powerful computer language that can be utilized to write source code, consisting of instructions and statements in a programming language. Therefore, the choice of language should be made with caution, as it directly impacts the implementation of equations, modules, and algorithms, while minimizing error standard deviation. The selection of programming languages for different models varies based on computational requirements and the nature of the problem at hand. Here are some noteworthy programming languages:

1) C: C is an imperative language that utilizes a relatively direct compiler and supports cross-platform coding. It is well-suited for recursion, structured programming, and lexical scoping. C offers a wide range of logical and arithmetic operators but has a limited number of keywords.
2) C++: C++ encompasses object-oriented, generic, and imperative programming features. It includes operators for logical operations, basic arithmetic, indirection, and more. C++ also provides static and dynamic polymorphism.
3) C#: C# is an object-oriented, general-purpose language that finds suitability in applications for embedded and hosted systems.
4) FORTRAN: FORTRAN is a dominant imperative language particularly suitable for scientific computing, such as weather

predictions, fluid dynamics, finite element analysis, and other related fields.

5) Java: Java is a class-based, generic, and imperative programming language. It is designed to minimize dependencies, allowing code to be applied without recompilation on any platform that supports Java.
6) MATLAB: MATLAB is well-suited for matrix operations and algorithms, and it facilitates data plotting.
7) Python: Python is an open-source software with readily available interpreters for a wide range of operating systems. This language promotes coding in fewer lines and emphasizes code readability.
8) R: R is an interpreted language primarily used for statistical computing, data analysis, and the development of statistical software. It offers numerous capabilities, including user-created packages.
9) VB.net: VB.net is an object-oriented language that includes a command-line compiler. It has incorporated several features through various language versions, such as partial classes, nullable types, anonymous types, lambda extensions, iterators, and more.

3. Role of GIS in Modeling

The Geographic Information System (GIS) is a widely used scientific tool among researchers today. It serves as a valuable framework for managing spatial data, including maps and data related to geographical conditions. These maps are connected to geo-referenced datasets, commonly referred to as Relational Database Management Systems (RDBMS). The datasets typically consist of raster and vector data, which are utilized within GIS tools.

The GIS framework effectively handles the organization of spatial data within a given study area. Additionally, it incorporates model-enabling methods that allow users to seamlessly integrate models within the GIS framework. This integration enables users to operate with the same GIS framework while working with the model. Consequently, essential data such as the length, width, or cross-section of a river can be easily obtained using GIS tools. Furthermore, GIS facilitates the calculation of catchment or basin areas for rivers or other study areas. Moreover, GIS tools are capable of swiftly extracting information like the slope of an area or the elevation of a specific location.

These measurements, including length, width, elevation, area, slope, and others, are numerical values that hold significance in computer programs. They are commonly used in computer models to process data synthesis, which involves converting raw data into a format that is easy to comprehend and analyze. To illustrate this concept, we can represent the workflow as follows:

GIS (geographical data) ⇒ MODEL ⇒ OUTPUT

4. Use of Graphical User Interface (GUI)

Considerable research has been conducted on integrated environmental modeling frameworks for environmental impact assessment (EIA) and decision support, particularly in the fields of agriculture and hydrology. Various authors have utilized separate frameworks for integrating models (Newham et al., 2004; Watson & Rahman, 2004), but this approach has proven to be burdensome and requires additional training, discouraging users from adopting it. Consequently, single-integrated models have garnered more attention due to their popularity and ease of use.

In the early 1990s, efforts were primarily focused on developing graphical user interfaces (GUIs) and simulation model frameworks with geographic information systems (GIS). The utilization of GIS facilitated the integration of extensive spatial data into a spatial database. Simulation models were subsequently integrated into GIS-GUI, enabling the assimilation of both spatial and non-spatial data. This integration of model simulations marked a significant advancement and was embraced by many researchers as a step forward in their work (Arnold et al., 2009).

While some authors have employed models that require manual interaction between the model and data (Muth & Bryden, 2013; Yeh et al., 2006), single-integrated models that operate independently as standalone units are more prevalent and typically preferred over other model forms. Single integrated models offer advantages over manual approaches, as they reduce the likelihood of human errors and facilitate easier data handling, including pre/post-processing of spatially distributed input data, as well as improved simulation capabilities.

Taking into account the aforementioned factors, such as the working principle, agro-hydrological modeling, and GUI, this chapter focuses on describing the SWAT model.

5. Some Common Environmental CSMs

There are numerous models commonly used by researchers nowadays, including various crop growth models such as ALMANAC, APSIM, AQUACROP, APEX, DSSAT, EPIC, STICS, CROPWAT, CROPSYST, CERES, INFOCROP, WEAP, among others, which are highly favored by agro-environmentalists. However, for the explanation, this study focuses on the Soil and Water Assessment Tool (SWAT) model. SWAT is an agro-hydrological, physically-based, scale model extensively employed by researchers worldwide for climate change and environmental studies.

5.1. Agro-Hydrological Models

In agro-hydrological modeling, the scale of simulations plays a crucial role. Typically, the model operates independently at different scales, including region, basin, catchment, sub-catchment, watershed, and micro-watershed. Consequently, models are classified into various sub-categories such as watershed or farm scale, surface water or groundwater, agro-hydrologic or farm economic models, and others. The combined requirements of agricultural, watershed, farm scale, economic, and hydro-climatic models have necessitated the use of different models to achieve the desired solutions.

Crop simulation models, commonly referred to as crop growth models (CGMs), offer a means to assess different farm management methods without the need for costly and time-consuming experiments. Most cropping system models are specific to particular sites and consist of extensive databases that are spatially and geo-referenced. One viable approach is the development of a database management system (DBMS) capable of providing both static and dynamic access to large volumes of geo-referenced data required by such applications (Yang et al., 2011). Modeling biophysical processes is a complex task that relies on extensive datasets, which introduces uncertainty in the model parameters (Wang et al., 2005).

5.2. The SWAT Model

Considerable attention has been focused on the evaluation of the SWAT model, including its model description, history, functioning, worldwide application, case studies, comparisons, strengths, weaknesses, and development since its inception (Gassman et al., 2010; Arnold et al., 2009). Over time, the SWAT model has gradually gained preference among environmental and hydrological researchers worldwide in various fields of study.

The SWAT (Soil and Water Assessment Tool; Arnold et al., 2012) model is capable of hydrological modeling, simulating various aspects of the hydrological balance in quantitative terms. Major components of the SWAT model include hydrology, weather parameters (such as temperature, rainfall, humidity, solar radiation, and wind velocity), sediment transport, crop growth (biomass), soil and air temperature, pesticides, nutrients, and agricultural management options (Arnold et al., 2009; Gassman et al., 2009; Nelson et al., 2006).

The SWAT model is integrated with ArcGIS-GUI, as shown in Figure 2, which facilitates data input and manipulation in the required format for the model. Additionally, it assists in executing the model and provides the model output in the desired standard format, such as MDB and text files. The front-end and back-end operations are depicted in Figure 2.

The key back-end procedure involves running the model, which entails executing the SWAT project. This process utilizes the model program (swat.exe), written in the FORTRAN (FORmula TRANslation) language. The program performs well in numerical computation, supported by the robust FORTRAN mathematical library. It reads data such as geographical area, stream length, and catchment slope, and carries out a multitude of calculations (over a thousand equations). The results are then stored in the output folder in text format, organized by watershed. The user interface accesses specific outputs for further analysis.

The SWAT model requires input data in a standard prescribed format, including raster data such as DEM, land use, and cadastral information, as well as vector data like stream networks and watershed boundaries. Additionally, it relies on a database that includes default parameters such as soil, crop, machinery, tillage, weather station details (including wind speed and direction), fertilizers, pesticides, and others. The weather station information consists of time series data, which can be seen in Figure 3.

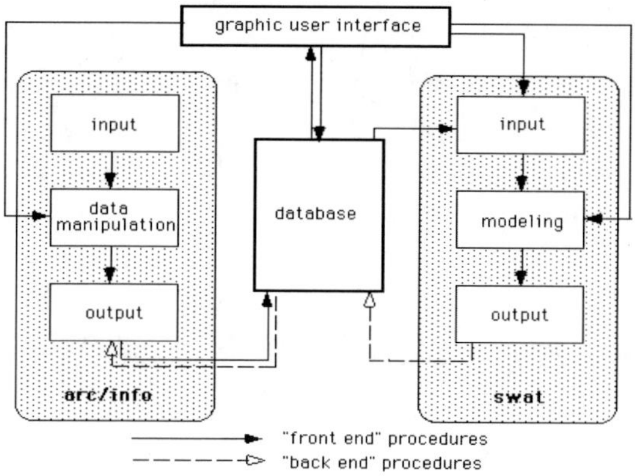

Figure 2. The architecture of the interface system coupling ArcGIS and SWAT (adopted from Bian et al., 1996).

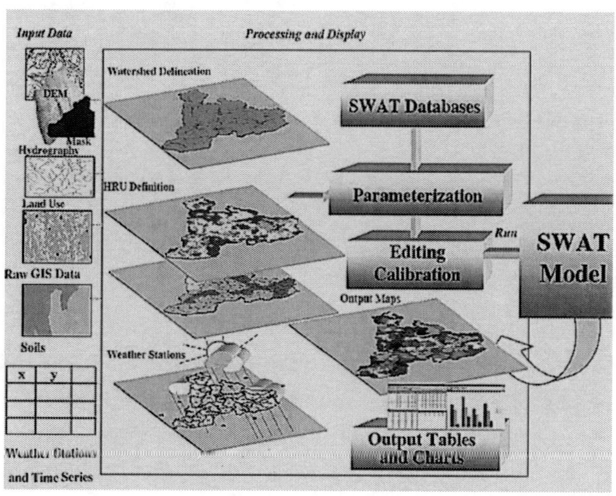

Figure 3. Schematic diagram of the GIS with the SWAT model (adopted from Di Luzio et al., 2002).

Once the data is provided as input, it undergoes processing through the ArcGIS backend. This process involves fetching data from the SWAT model project database and executing the model, as shown in Figure 4. Subsequently, the model generates results and outputs in a prescribed dataset/format. The SWAT model interface and its working approach are briefly described in Figure 4.

The SWAT model is a watershed-scale simulation model that replicates a range of physical processes. It enables users to validate data by comparing it to observed values with statistical confidence. Furthermore, the model offers the ability to assess and quantify the "climatic impacts" in intricate watersheds characterized by diverse soil types, management practices, and land use conditions over extended periods of time. These impacts are commonly referred to as "climate change" by the World Meteorological Organization (WMO).

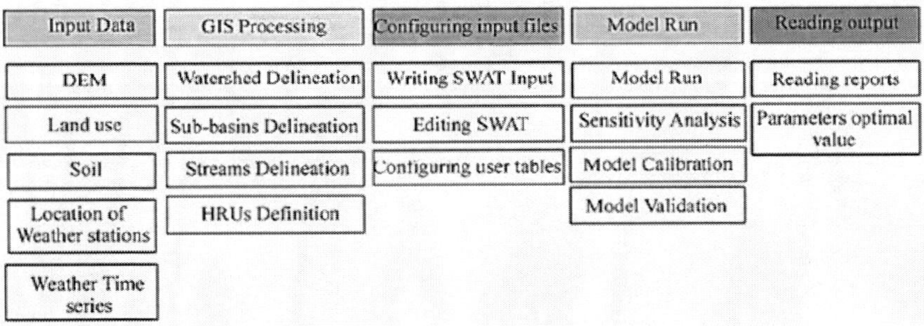

Figure 4. Components and input/output data of SWAT model (adopted from Fadil et al., 2011).

6. Climate Change Studies

Climate change studies are a captivating and complex endeavor due to the involvement of various parameters (physical, chemical, and biological) that interact simultaneously. These parameters exhibit spatial and temporal variations, prompting the development of numerous scientific models over the past fifty years across different research fields. This progress has been facilitated by the wide array of methods, techniques, and theories available in science, catering to specialized domains.

Scientific computer-based modeling, encompassing conceptual, graphical, and mathematical models, plays a crucial role in scientific activities. Simulation modeling has become an integral component of nearly every scientific endeavor. Different technological disciplines have their own approaches to specific types of modeling. However, there is a lack of general theory on scientific modeling, including knowledge visualization (Arnold et al., 2009).

Among researchers, two types of models are particularly popular: global climate models (GCMs) and regional climate models (RCMs). These models provide climate data, which, in turn, are utilized in CGMs to simulate and examine the agro-hydrological balance of a given area.

6.1. Farm Advisory Services

It is well-established that farming in India, particularly for small-scale and marginal farmers, has always been a demanding task. To alleviate the challenges associated with farming, it is crucial to embrace modern techniques supported by computer modeling systems. By employing modeling approaches in agriculture, farmers are afforded the opportunity to effectively manage resources and make informed decisions. Adopting a realistic modeling approach enables optimal utilization of farm resources, leading to the attainment of maximum crop yield. Moreover, it offers a means to address significant constraints, such as chemical and irrigation water requirements, based on rainfall patterns, crop evapotranspiration, and the availability of other resources.

6.2. Constraints of the Environmental Models

1) The utilization of models involves processing large amounts of data through lengthy computer programs. These data are collected and processed at various locations, employing diverse collection methods that introduce inherent errors. Consequently, managing data errors becomes challenging due to the diverse nature of data sources, leading to concerns regarding the uncertainty and accuracy of the data.
2) The scale of modeling and the resolution of data significantly impact the resulting output. Higher data resolution yields better model results. However, the weather data used during the study becomes less precise over time, while high-resolution data is becoming available and is expected to become even more refined in the future. Therefore, the evolution of data refinement over time also contributes to the overall uncertainty and accuracy.
3) Numerous assumptions, such as consistent inflation rates and weather scenarios following the same trend for the specified time period, can introduce inaccuracies in computations. The primary sources of potential errors include model-based assumptions and

additional errors arising from price prediction and calculation based on inflation. The study itself relies on multiple assumptions regarding parameters and built-in equations. Price prediction, in particular, depends on crop yield, which is a result of simulation and propagates simulation errors.

4) Other critical assumptions include constant land use, soil conditions, the absence of modern mechanization, and the exclusion of future development projects, among others. The presence of numerous assumptions raises concerns about the reliability of predictions.
5) Climate change modeling, particularly related to agricultural impact studies, often neglects the assimilation of carbon dioxide, which is considered an aerial fertilizer by many researchers. This omission can significantly influence climate change outcomes. In most studies, carbon dioxide is assumed to remain constant, although in reality, it is expected to vary. However, due to the lack of available data series, it is typically treated as a constant for the specified study period.
6) The growth and distribution of pests and crop diseases are highly influenced by air and soil temperature, rainfall patterns and amounts, humidity, and the length of seasons. However, these crucial aspects are often overlooked in climate change studies.

6.3. Way Forward to Environmental Models

In general, agriculture and the environment, including ecosystems, are influenced by various factors such as carbon dioxide concentration, temperature, solar radiation, precipitation, wind velocity, and air humidity. These factors interact with each other and are described by multiple parameters and constants in equations. This implies that the interactions between biotic and abiotic factors are complex and extensive. Consequently, the scale of ecosystems is too vast to be fully captured by a few thousand equations in a computer model created by humans. Nevertheless, scientists employ a comprehensive and organized mathematical and scientific approach to draw conclusions and make decisions based on sound scientific reasoning.

Conclusion

Computer models are essential in the present era, where the processing of large amounts of data is made easy by powerful machines. These models enable the integration and coupling of multiple data sources, where the output of one model serves as input for another. This approach allows for manageable scaling of modeling efforts, which is crucial for ensuring accuracy and managing uncertainty. Computer models have proven valuable in various sectors, including agriculture, surface water management, groundwater analysis, forest conservation, and oceanic studies. They have also played a significant role in climate change research, gene mapping, photosynthesis studies, and investigating air and water pollution, among other applications.

References

Allen, M. P., & Tildesley, D. J. (1987). *Computer simulation of liquids.* https://web.archive.org/web/20140903110924/ http://www2.chemistry.msu.edu/courses/cem888/cukier/Allen,Tildesley,_Computer_Simulation_of_Liquids,1991.pdf. Oxford University Press.

Arnold, J., Srinivasan, R., Neitsch, S., George, C., Abbaspour, K., Gassman, P., Hao, F. H., Griensven, A. V., Gosain, A. K., Debels, P., Kim, N. W., Somura, H., Ella, V., Leon, L., Jintrawet, A., Reyes, M., & Sombatpanit, S. (2009). Soil and Water Assessment Tool (SWAT) global applications. World Association of Soil and Water Conservation Special Publication no. 4. *Bangkok, Thailand. 397 p.p.*. ISBN 978-974-613-722-5.

Arnold, J. G., Moriasi, D. N., Gassman, P. W., Abbaspour, K. C., White, M. J. Srinivasan, R., Santhi, C., Harmel, R. D., van Griensven, A., van Liew, M. W., Kannan, N., & Jha., M. K. (2012). SWAT: Model use, calibration, and validation. *Transactions of the ASABE*, *55*(4), 1491–1508. https://doi.org/10.13031/2013.42256.

Bian, L., Sun, H., Blodgett, C., Egbert, S., Li, W., Ran, L., & Koussis, A. (1996). An integrated interface system to couple the SWAT model and Arc/Info. *Proceedings, Third International Conference on Integrating Geographic Information Systems and Environmental Modeling.* https://web.archive.org/web/20100626210516/ http://www.ncgia.ucsb.edu/conf/SANTA_FE_CD-ROM/sf_papers/bian_ling/lbian.html.

Di Luzio, M., Srinivasan, R., & Arnold, J. G. (2002). Integration of watershed tools and SWAT model into BASINS. *Journal of the American Water Resources Association*, *38*(4), 1127–1141. https://doi.org/10.1111/j.1752-1688.2002.tb05551.x.

Fadil, A., Rhinane, H., Kaoukaya, A., Kharchaf, Y., & Bachir, O. A. (2011). Hydrologic modeling of the Bouregreg watershed (morocco) using GIS and SWAT model.

Journal of Geographic Information System, 03(4), 279–289. http://doi.org/10.4236/jgis.2011.34024.

Gassman, P. W., Williams, J. R., Wang, S., Saleh, A., Osei, E., Hauck, L., Izaurralde, C., & Flowers, J. (2009). The Agricultural Policy Environmental Extender (APEX) model: An emerging tool for landscape and watershed environmental analyses. Technical Report 09. https://web.archive.org/web/20140505172722/http://www.card.iastate.edu/publications/dbs/pdffiles/09tr49.pdf. *Technical Report* 09-TR 49. *Center for Agricultural and Rural Development*. Iowa State University.

Gassman, P. W., Williams, J. R., Wang, X., Saleh, A., Osei, E., Hauck, L. M., Izaurralde, R. C., & Flowers, J. D. (2010). Policy environmental EXtender (APEX) model: An emerging tool for landscape and watershed environmental analyses. *Transactions of the ASABE, 53*(3), 711–740.

Muleta, M. K., & Nicklow, J. W. (2005). Decision support for watershed management using evolutionary algorithms. *Journal of Water Resources Planning and Management, 131*(1), 35–44. https://doi.org/10.1061/(ASCE)0733-9496(2005)131:1(35).

Muth, D. J., & Bryden, K. M. (2013). An integrated model for assessment of sustainable agricultural residue removal limits for bioenergy systems. *Environmental Modelling and Software, 39*, 50–69. https://doi.org/10.1016/j.envsoft.2012.04.006.

Nelson, R. G., Ascough, J. C., & Langemeier, M. R. (2006). Environmental and economic analysis of switchgrass production for water quality improvement in northeast Kansas. *Journal of Environmental Management, 79*(4), 336–347. https://doi.org/10.1016/j.jenvman.2005.07.013.

Newham, L. T. H., Letcher, R. A., Jakeman, A. J., & Kobayashi, T. (2004). A framework for integrated hydrologic, sediment and nutrient export modelling for catchment-scale management. *Environmental Modelling and Software, 19*(11), 1029–1038. https://doi.org/10.1016/j.envsoft.2003.11.006.

Wang, X., He, X., Williams, J. R., Izaurralde, R. C., & Atwood, J. D. (2005). Sensitivity and uncertainty analyses of crop yields and soil organic carbon simulated with EPIC. *Transactions of the ASAE, 48*(3), 1041–1054. http://dx.doi.org/10.13031/2013.18515.

Watson, F. G. R., & Rahman, J. M. (2004). Tarsier: A practical software framework for model development, testing and deployment. *Environmental Modelling and Software, 19*(3), 245–260. https://doi.org/10.1016/S1364-8152(03)00152-X.

Yang, Y., Wilson, L. T., Wang, J., & Li, X. (2011). Development of an integrated cropland and soil data management system for cropping system applications. *Computers and Electronics in Agriculture, 76*(1), 105–118. https://doi.org/10.1016/j.compag.2011.01.012.

Yeh, S., Wang, C., & Yu, H. (2006). Simulation of soil erosion and nutrient impact using an integrated system dynamics model in a watershed in Taiwan. *Environmental Modelling and Software, 21*(7), 937–948. https://doi.org/10.1016/j.envsoft.2005.04.005.

Chapter 13

An Analysis of Bioenergy Plantations for the Reclamation of Degraded Lands and Their Sustainability

Saurabh Singh*
and Jay Prakash Verma
Institute of Environment and Sustainable Development,
Banaras Hindu University, Varanasi, India

Abstract

There are many different ways to use the word "restoration," but it often refers to going back to something's initial state. It should be viewed as encompassing entire ecosystems in ecological restoration. It is important to keep in mind that there are restoration alternatives that can be more realistic, such as replacement or rehabilitation. Recently, land degradation problems have greatly come into the picture because of the increased anthropogenic activities. To ensure the land isn't wasted and utilized for meaningful purposes, it is important to restore the land using sustainable options. Bioenergy plantations with their wide range of adaptability and ability to grow in stressed conditions provide the option for the restoration of degraded lands. Natural restorative mechanisms offer many beneficial aspects which is why they should be used whenever and wherever possible. However, each of them may require a specific treatment plan because restoration is an ongoing process, and specifying success criteria can be challenging. Different bioenergy plantations such as Jatropha, miscanthus, and

* Corresponding Author's Email: srb0484@gmail.com.

In: Conserving, Halting and Reinstating Degraded Natural Resources …
Editors: Rajani Srivastava and Pankaj Kumar
ISBN: 979-8-89113-067-8

switchgrass can be used for such purposes. This chapter discusses land degradation, its causes and restoration using bioenergy plantations.

Keywords: ecological restoration, rehabilitation, recuperation, reclamation

1. Introduction

The increasing population and their ever-increasing energy demands are compelling researchers to explore new methods for energy conversion (Dincer, 2000). The increasing world population due to the changes in land use and variation in the land cover of plantations has led to an increase in the area of degraded lands (Alemu, 2015). Along with the increasing population, an increase in the energy demands of the people has also been observed which further prompts new methods of energy conversion. Land degradation issues have seen a sudden surge in the recent past due to anthropogenic activities resulting from the increased world population, mainly increasing the cropland and grazing land for more crops and cattle respectively (Hossain et al., 2020). The increasing world population has affected every single natural resource available on Earth and land isn't an exception to it. The extent and rate of degradation of land have become a major problem owing to its overexploitation and also the extraction of resources. The increasing population puts pressure on land resources as demand for housing, agriculture, and other uses of land increases, leading to overdevelopment, deforestation, loss of biodiversity, and degradation of ecosystems (Eswaran et al., 2019; Hossain et al., 2020; Mohsin et al., 2019). In the year 2010-11, 4.1 m ha of wasteland that could be used for agricultural activities was recorded. Further with the increasing use of fossil fuels high amounts of emissions are being released into the atmosphere, which is why biofuel production is emphasized. Biofuel production from bioenergy plants can help in many ways to curb the usage of fossil fuels (Singh et al., 2022). These bioenergy plants can be used for the production of biodiesel as well as bioethanol. A lot of research is underway to produce high yields of second-generation biofuel due to the competitive nature of first-generation biofuel feedstock with food crops but if the bioenergy plants such as jatropha, miscanthus and switchgrass are used on degraded lands that can be beneficial in both ways (Morya et al., 2022; Singh et al., 2020a, 2019, 2018; Singh & Verma, 2019).

With the increasing demand for food, the overexploitation of land for agricultural purposes has been done which negatively affects the quality of soil on land. Bad agriculture practices are the major factor in the degradation of soil. It can lead to soil degradation which may be induced due to excessive tillage, monoculture crops, and intensive use of pesticides and fertilizers thereby reducing soil fertility and leading to soil erosion, making the land less productive in the long term (Doran et al., 2018). Further, biodiversity loss occurs due to clearing large areas of land for agriculture which destroys habitats and consequently the loss of plant and animal species (Okia, 2012). Further, it leads to microbial flora decline, which significantly impacts soil health and productivity (Knops et al., 2002). Some microbes involved in decomposing organic matter and breaking down toxic compounds suppress plant diseases, and their loss can make the soil more vulnerable to degradation and diseases (Bailey & Lazarovits, 2003). Also, increased pesticide and chemical fertilizers used in agriculture make it even more difficult to maintain soil quality and microbial flora (Gouda et al., 2018). Pesticides and fertilizers alter the chemical composition of the soil thereby reducing fertility and hence soil degradation (Tripathi et al., 2020). Some chemicals such as heavy metals, salts, and some organic compounds, bind to soil particles making up essential nutrients, viz., nitrogen, phosphorus, and potassium, unavailable to plants, consequently reducing soil fertility and declining crop productivity.

Further, land degradation can also be induced due to climate change as variations in temperature and moisture directly affect the soil quality and its microbial flora (Cowie et al., 2011). Reduction in the moisture and reduced binding strength of soil due to reduced plantations resulting from a reduction in microbial flora and hence soil fertility may further lead to erosion, landslides, and further decreased water retention (Kurganova et al., 2022; Mills & Fey, 2004). Deforestation for construction, as well as agriculture practice, is another problem which leads to soil degradation. When forest soils are exposed after clearing of forests it leads to an exponential increase in erosion and runoff in soil which reduces its fertility and ability to retain water (Zuazo & Pleguezuelo, 2009). Likewise, there are many such factors associated with the land degradation problem and this chapter discusses most of them and their remedial measures using bioenergy plantations. A sustainable solution for its mitigation can be the bioenergy plantations which have shown immense potential for the restoration of degraded lands due to their morphological as well as physiological characteristics.

2. Causes of Land Degradation

The main causes of land degradation can vary from political to social to industrial as well as economic aspects (Singh et al., 2020b). The land degradation effects might not be instantly visible from the activity and may take time due to which they often go unnoticed until it's too late for recovery. This often results because of the lack of awareness by the local groups and people residing in the area. For example, the indiscriminate use of chemical fertilizers in an agricultural field for the production of crops to increase overall productivity often results in the decreased vigour and fertility of the soils with time (Faroque et al., 2011; Thakur, 2017). When the chemicals are used for increasing productivity the farmers generally tend to ignore the negative effects produced by these chemicals on the soil and focus directly on the crop yield. The increase in yield is directly linked to the chemicals applied to the field (Zingore et al., 2008). The residual amounts of these chemicals in the field accumulate to hazardous levels which in the long run results in the degradation of the soil and land.

Another example of land degradation is the effluents from industries which contain toxic chemicals and are often released into the nearby water bodies and fields without proper treatment (Aulakh & Sidhu, 2015). This can result in the accumulation of toxic chemicals in the fields and can directly affect the fertility of soils in terms of their nutrients as well the beneficial microbial flora present in the soil (Bharti et al., 2017; Wong, 2003). The toxic chemicals may include heavy metals such as chromium (Su et al., 2016), cadmium (Ma et al., 2020), lead, arsenic and many other effluents that are often released from industries (Gan et al., 2009). The accumulation of chemicals in soils adjacent to the industries results in the decrease of the beneficial microbial flora and also results in the decline of the vegetation consequently. The decline in vegetation is linked to the increased erosion of the soil. The increased erosion leads to the loss of the top fertile layer of the soil and thus the process of land degradation starts.

A major factor responsible for the degradation of the land is overgrazing by cattle (Kairis et al., 2015). There are many areas where people tend to allow their cattle on grass that is present in a particular area that ultimately declines to a level which indues erosion of soil (Aulakh & Sidhu, 2015). The erosion of soil then drives the process of land degradation as described earlier. The overgrazing of lands for cattle feeds can be regarded as the major factor responsible for the land degradation process which is often driven by economic force. The cattle grazing and cropland utilization for small scales does not necessarily instigate the process of land degradation, but when these processes are driven by industrial demands they can lead to enhanced levels of grazing and cropland utilization which then initiates the process of erosion, decreased nutrient level, low water holding capacity and ultimately the degradation of soil and land.

Another factor that can be held responsible for the degradation of the land is the changing lifestyle patterns which are often linked to higher food requirements, and higher dairy product requirements which are ultimately linked to land degradation. Moreover, many other industrial and economic factors often go unnoticed concerning land degradation. Until and unless the overall scenario and factors linked directly and indirectly to land degradation are assessed properly and suitable policies to overcome aren't framed, the process of land degradation will continue. Proper agricultural practices and sustainable management of farmlands followed by suitable measures taken at the policymaking level might help curb the land degradation process and the impacts associated with it.

3. Bioenergy Plantations

Bioenergy plantations have been used for the production of biofuels in the form of biodiesel production for a long time, for example from Jatropha plants (Koh & Tinia, 2011). But their use due to the first-generation biofuel production got restricted because of their competitive nature with the food crops if grown on land that can be used for cultivation of food crops. However, bioenergy plants due to their physiological adaptations offer great prospects in terms of survivability under low nutrient as well as stress conditions (da Costa et al., 2019; Karp & Shield, 2008). This quality of the bioenergy plants can be utilized for the maintenance of land that has been degraded by growing them in such conditions and instigating and reinvigorating the natural processes necessary for a healthy land. Bioenergy crops are fast-growing trees, grasses, and other plants which can be harvested and later on processed into biofuels, including ethanol and biodiesel, as well as other forms of renewable energy, such as biogas. Bioenergy plantations provide a sustainable energy source which can reduce fossil fuel dependency and simultaneously mitigate the impacts of climate change. One major advantage of bioenergy plantations is that they can be established on land unsuitable for food crops, such as marginal land, abandoned land, and degraded land which helps reduce competition for fertile land and prevent deforestation. Bioenergy crops can also provide other benefits, such as soil conservation and improvement, and job creation in rural areas through increased crop production and their associated activities.

One of the major potential bioenergy crops is Miscanthus, a perennial rhizomatous grass which grows up to 10 feet in height and produces high yields of biomass including varieties yielding more than 20 dry tons per acre per year. It can be harvested annually and processed into various forms of bioenergy, including solid biofuels, such as pellets and briquettes, and liquid biofuels, such as ethanol (Chung & Kim, 2012). Also, miscanthus offers many other pros as a bioenergy crop such as low maintenance because it requires minimal inputs of fertilizers and pesticides and is also highly resilient to pests and diseases (Brosse et al., 2012). Additionally, miscanthus' deep root system helps prevent soil erosion and improve soil health, further making it beneficial for marginal and degraded land (Kowalczyk-juśko et al., 2022).

Salt-affected soil is spread across many parts of the world and is estimated to be affecting nearly 20% of the world's irrigated land. The development of crops that can be grown in salt-affected areas can reduce the negative impact of salinity on food production as well as utilize the initially unused land. By developing crops that can grow in these areas, it may be possible to reduce the negative impacts of salinity on food production and increase the land available for bioenergy production. Common examples include switchgrass, miscanthus, and poplar, as potential feedstocks for the production of biofuels grown in salinity stress conditions (Table 1) (Buttar et al., 2020; Ye et al., 2010). The various mechanisms in plants that resist salinity stress are improved ion exclusion, enhanced osmoregulation, anti-oxidant defense, increased root growth, high-stress tolerance, and also reduced transpiration rate (Hangs et al., 2011; Li et al., 2014; Mu & Ding, 2022). Drought-resistant bioenergy plants are also used which are genetically modified or selectively bred to withstand drought conditions and maintain productivity in water-scarce conditions (Oliver et al., 2009). These crops developed as potential feedstocks for the production of biofuels owing to their potential to provide a sustainable source of energy while reducing the negative impact of water scarcity on agriculture. An estimated 30% of the world's cropland is affected by drought, the development of drought-resistant bioenergy crops becomes more important because of its potential to increase land reclamation for bioenergy production and other useful purposes. The mechanism of drought tolerance by the bioenergy crops includes improved water use efficiency, increased root depth, enhanced stomatal regulation and osmotic regulation, and reduced transpiration rate.

Table 1. Bioenergy plantations for the restoration of contaminated lands

Name of bioenergy plant	Type of degradation of land	Effects or mitigation impacts	Reference
Miscanthus Giganteus	Degraded Coal Mine Soils	Increase in plant biomass observed with increasing years upon addition of fertilizers.	(Jeżowski et al., 2017)
Miscanthus Giganteus	Soils polluted by metal(loid)s	Proper development of the plant with sufficient biomass production. Sequestration of metals by the plants.	(Pidlisnyuk et al., 2020)
Jatropha curcas L.	highly contaminated mining soils	J. curcas L. absorbed Fe (>3000 mg kg^{-1} plant), Pb, Zn, Cu, Cr and Ni, and traces of As.	(Álvarez-Mateos et al., 2019)
Jatropha curcas L.	soil contaminated with used lubricating oil	Jatropha root did not sequester hydrocarbons but helped in degradation through rhizodegradation mediated by microbes	(Agamuthu et al., 2010)
Panicum virgatum	Heavy Metal-Contaminated Soi	Amendments with thermophilic compost, vermicompost, and coconut coir increased the plant biomass and decreased the amount of bioavailable metals in the soils	(Shrestha et al., 2019)

Another quality, i.e., flood tolerance of bioenergy plantations can be utilized for growth in degraded lands flooded with water. Examples of flood-tolerant bioenergy plantations include Miscanthus, Eucalyptus, Pinus, Poplus, and Sorghum (Leontopoulos & Arabatzis, 2021). The major advantage associated with flood-tolerant bioenergy plantations is that their overall yield in general is higher which in turn may reflect in their overall biofuel yield. Different mechanisms of increased flood tolerance by bioenergy plants include improved root aeration, enhance anti-oxidant defense mechanism, increased shoot length and enhance nutrient uptake (Schuler et al., 2009).

Many bioenergy crops such as certain species of legumes (e.g., soybeans, alfalfa, clover) help in the restoration of degraded lands by increasing the nitrogen content in the soil. This is achieved through the association of nitrogen-fixing microbes with the roots of these plants. Leguminous crops are such plants which can help in the fixation of atmospheric nitrogen thereby increasing the fertility of soils. In this, there is a symbiotic relationship between the bacteria and plants wherein the bacteria form nodules on the roots of the plant, where they convert atmospheric nitrogen into a form of nitrogen that the plant can use. The plant provides the bacteria with a source of energy (carbohydrates) in exchange for nitrogen. This allows the plant to grow and thrive without the need for added nitrogen fertilizer.

Miscanthus and switchgrass have been shown to sequester cadmium (Cd) from the soil (Guo et al., 2016; Reed et al., 2006). Studies have shown that both these bioenergy plants have a high capacity for cadmium sequestration, with the ability to absorb and store large amounts of the metal in their roots and shoots. This makes it an effective tool for reducing the levels of cadmium in contaminated soils. The mechanism of cadmium sequestration in switchgrass is thought to involve the uptake of cadmium ions into the root cells, followed by translocation of the metal to the shoots and storage in the plant tissues (Afzal et al., 2017). Switchgrass has also been shown to possess high levels of metal-chelating compounds, which can further enhance its ability to sequester cadmium. Studies have also shown that crops such as switchgrass, Miscanthus, and willow (*Salix* spp.) can sequester chromium and arsenic from the soil (Arduini et al., 2006; C. Li et al., 2011; Reed et al., 2013; Yanitch et al., 2017; Yu et al., 2008). The overall mechanism of heavy metal stress tolerance by bioenergy plants is similar which is generally through sequestration and phytostabilization. However, arsenic tolerance involves a combination of factors such as the ability to produce arsenic-chelating compounds, exclusion of arsenic from critical cellular processes, and also maintenance of normal cellular functions even in the presence of toxic metal. Roots of bioenergy crops create a microhabitat for microbes by providing a physical structure for them to colonize which helps in promoting the growth of beneficial microorganisms, such as nitrogen-fixing bacteria essential for soil health and fertility. Many studies testify to the fact that bioenergy crops such as switchgrass and Miscanthus, increase the abundance of several important groups of soil microbes, including fungi, bacteria, and actinomycetes. These changes in the soil microbial community positively impact soil health, nutrient cycling, and consequently plant growth and also the carbon and nutrient sequestration potential of the soil.

4. Transgenic Plants

Transgenic bioenergy plantations offer great prospects for the restoration of degraded lands (Singh et al., 2020b). Transgenic bioenergy plants can be modified according to the requirements of a specific type of degraded land

and can be used for restoration. Many genetically engineered bioenergy plants can remove certain toxic chemicals from the soil, increase the nitrogen content in the soil, and help increase the beneficial microbial flora by facilitating processes such as phytostabilization and rhizofiltration (Nedjimi, 2021). Genetically engineered bioenergy plants might also have the enhanced ability to resist salinity stress, water stress, drought stress and various other types of biotic and abiotic stresses which may be associated with degraded land. Genetically engineered plants in the recent past have made huge progress towards the same and are employed mainly for heavy metal detoxification, organic pollutant degradation and also overcoming salinity and drought stress. Various studies for the removal of heavy metals induced the overexpression of genes responsible for metallothioneins, and phytochelatins followed by increased organic pollutant sequestration by the plants (Memon et al., 2001; Yang et al., 2005). However, there are some concerns associated with transgenic bioenergy plants, such as the genetic cross-contamination potential with non-transgenic crops, and the potential for negative effects on the environment and also human health. The ongoing debate about the acceptability of genetically engineered plants has become a major focus of research and public discussion, which is ultimately dependent upon several factors, including the perceived benefits and risks, the available scientific evidence, and the specific context in which the technology is being used (Singh et al., 2020b).

5. Sustainability of Bioenergy Plantations for Restoration of Degraded Lands

Bioenergy plantations having the potential to play a role in degraded land restoration cannot mask the potential risks associated with these plantations and a careful evaluation of both risks and benefits of these plantations is necessary, followed by their implementation in a sustainable manner. For this, a comprehensive sustainability analysis that takes into account the impacts of bioenergy plantations on the environment, local communities, and the global economy is required. The sustainability of the bioenergy plantations for the restoration of degraded lands depends on many factors such as the type of crop used, the method of cultivation, and the management practices employed. The type of bioenergy crop is crucial for the sustainability of the plantation. Some crops, such as willow and poplar, are fast-growing and have low maintenance requirements, making them well suited for degraded land restoration. Other crops, such as switchgrass, are more drought tolerant and can provide a valuable source of feedstock for bioenergy production. Further, transgenic plants such as miscanthus, switchgrass, jatropha and poplar can enhance restoration by increasing physiological functions introduced into them during bioengineering. Certain cultivation methods such as agroforestry and conservation agriculture have shown promising results concerning sustainability. Further using bioenergy crops for their growth on degraded lands can help sequester carbon and in turn can help in the production of biofuels which greatly adds to the benefits towards sustainability. The production of biofuels from bioenergy plantations can help reduce the dependency on fossil fuels which further strengthens the sustainability aspect. The production of fuels from the bioenergy crops on degraded lands can on one hand increase the overall fertility of the soils which in turn will increase the biodiversity while on the other hand, they will help in mitigating the effects of fossil fuel burning by reducing their dependency. Further, the type of cultivation method of bioenergy plants depending upon the variety of plants will help mitigate land degradation problems. If proper measures are taken it can help reduce the erosion of soil on dry lands.

Conclusion

The degraded lands which are usually barren can be restored by increasing the organic carbon content of the soil through different methods. Further, fertility enhancements with respect to other nutrients can be brought about by choosing specific types of crops. The bioenergy plantations with high nutrient fixation in the surrounding soils may help enhance the fertility of soils and thereby restore the degraded lands. Bioenergy plantations can play a crucial role in the reclamation of degraded lands and contribute to sustainable development. These plantations can help restore degraded lands by increasing soil fertility and improving soil structure. The production of bioenergy from these plantations also provides a renewable source of energy and reduces dependence on non-renewable fossil fuels. There are a few drawbacks associated with the use of bioenergy plants for the restoration of barren lands but these can be mitigated using transgenic plants. There are many transgenic bioenergy plants with salt tolerance, drought tolerance toxic metal resistance and many other beneficial properties which help avoid the glitches associated with naturally occurring bioenergy plantations. However, considering the sustainability of these plantations from an environmental, social, and economic perspective becomes necessary to avoid any kind of drawbacks associated with bioenergy plantations. This may include appropriate species selection, using sustainable land use practices, and also fair distribution of benefits to local communities. Furthermore, the potential impacts of bioenergy production on food security and biodiversity should also be considered and addressed. With careful planning and implementation, bioenergy plantations can contribute to sustainable development and help in the restoration of degraded lands.

References

Afzal, S., Begum, N., Zhao, H., Fang, Z., Lou, L., & Cai, Q. (2017). Influence of endophytic root bacteria on the growth, cadmium tolerance and uptake of switchgrass (*Panicum virgatum* L.). *Journal of Applied Microbiology*, *123*(2), 498–510. https://doi.org/10.1111/JAM.13505.

Agamuthu, P., Abioye, O. P., & Aziz, A. A. (2010). Phytoremediation of soil contaminated with used lubricating oil using Jatropha curcas. *Journal of Hazardous Materials*, 179(1–3), 891–894. https://doi.org/10.1016/J.JHAZMAT.2010.03.088.

Akais Okia, C. (2012). Global Perspectives on Sustainable Forest Management. *BoD–Books on Demand.*

Alemu, B. (2015). *The Effect of Land Use Land Cover Change on Land Degradation in the Highlands of Ethiopia*. *5*(1).

Álvarez-Mateos, P., Alés-Álvarez, F. J., & García-Martín, J. F. (2019). Phytoremediation of highly contaminated mining soils by Jatropha curcas L. and production of catalytic carbons from the generated biomass. *Journal of Environmental Management,* 231, 886–895. https://doi.org/10.1016/J.JENVMAN.2018.10.052.

Arduini, I., Masoni, A., & Ercoli, L. (2006). Effects of high chromium applications on miscanthus during the period of maximum growth. *Environmental and Experimental Botany*, *58*(1–3), 234–243. https://doi.org/10.1016/j.envexpbot.2005.09.004.

Aulakh, M. S., & Sidhu, G. S. (2015). Soil Degradation in India: Causes, Major Threats, and Management Options. *MARCO Symposium.*

Bailey, K. L., & Lazarovits, G. (2003). Suppressing soil-borne diseases with residue management and organic amendments. *Soil and Tillage Research*, *72*(2), 169–180. https://doi.org/10.1016/S0167-1987(03)00086-2.

Bharti, V. S., Dotaniya, M. L., Shukla, S. P., & Yadav, V. K. (2017). Managing soil fertility through microbes: Prospects, challenges and future strategies. *Agro-Environmental Sustainability*, *1*, 81–111. https://doi.org/10.1007/978-3-319-49724-2_5/COVER.

Brosse, N., Dufour, A., Meng, X., Sun, Q., & Ragauskas, A. (2012). Miscanthus: a fast-growing crop for biofuels and chemicals production. *Biofuels, Bioproducts and Biorefining*, *6*(5), 580–598. https://doi.org/10.1002/BBB.1353.

Buttar, H. K., Badyal, R. K., Kumar, V., Singh, R. P., & Manchanda, G. (2020). Salt Stress Induced Morphological, *Anatomical and Ionic Alterations in Chickpea. 52*(6), 563–575. https://doi.org/10.1080/00103624.2020.1862155.

Chung, J.-H., & Kim, D.-S. (2012). Miscanthus as a potential bioenergy crop in East Asia. *Journal of Crop Science and Biotechnology 2012 15:2*, *15*(2), 65–77. https://doi.org/10.1007/S12892-012-0023-0.

Cowie, A. L., Penman, T. D., Gorissen, L., Winslow, M. D., Lehmann, J., Tyrrell, T. D., Twomlow, S., Wilkes, A., Lal, R., Jones, J. W., Paulsch, A., Kellner, K., & Akhtar-Schuster, M. (2011). Towards sustainable land management in the drylands: Scientific connections in monitoring and assessing dryland degradation, climate change and biodiversity. *Land Degradation & Development*, *22*(2), 248–260. https://doi.org/10.1002/LDR.1086.

da Costa, R. M. F., Simister, R., Roberts, L. A., Timms-Taravella, E., Cambler, A. B., Corke, F. M. K., Han, J., Ward, R. J., Buckeridge, M. S., Gomez, L. D., & Bosch, M. (2019). Nutrient and drought stress: implications for phenology and biomass quality in miscanthus. *Annals of Botany*, *124*(4), 553–566. https://doi.org/10.1093/AOB/MCY155.

Dincer, I. (2000). Renewable energy and sustainable development: a crucial review. *Renewable and Sustainable Energy Reviews*, *4*(2), 157–175. https://doi.org/10.1016/S1364-0321(99)00011-8.

Doran, J. W., Jones, A. J., Arshad, M. A., & Gilley, J. E. (2018). Determinants of Soil Quality and Health. *Soil Quality and Soil Erosion*, 17–36. https://doi.org/10.1201/9780203739266-2.

Eswaran, H., Lal, R., & Reich, P. F. (2019). Land degradation: An overview. *Response to Land Degradation*, 20–35. https://doi.org/10.1201/9780429187957-4.

Faroque, M. A. A., Kashem, M. A., & Bilkis, S. E. (2011). Sustainable Agriculture: A Challenge in Bangladesh. *International Journal of Agricultural Research, Innovation and Technology*, *1*(1–2), 1–8. https://doi.org/10.3329/ijarit.v1i1-2.13922.

Gan, S., Lau, E. v., & Ng, H. K. (2009). Remediation of soils contaminated with polycyclic aromatic hydrocarbons (PAHs). *Journal of Hazardous Materials*, *172*(2–3), 532–549. https://doi.org/10.1016/j.jhazmat.2009.07.118.

Gouda, S., Kerry, R. G., Das, G., Paramithiotis, S., Shin, H. S., & Patra, J. K. (2018). Revitalization of plant growth promoting rhizobacteria for sustainable development in agriculture. *Microbiological Research*, *206*, 131–140. https://doi.org/10.1016/j.micres.2017.08.016.

Guo, H., Hong, C., Xiao, M., Chen, X., Chen, H., Zheng, B., & Jiang, D. (2016). Real-time kinetics of cadmium transport and transcriptomic analysis in low cadmium accumulator Miscanthus sacchariflorus. *Planta*, *244*(6), 1289–1302. https://doi.org/10.1007/S00425-016-2578-3/figures/8.

Hangs, R. D., Schoenau, J. J., van Rees, K. C. J., & Steppuhn, H. (2011). Examining the salt tolerance of willow (Salix spp.) bioenergy species for use on salt-affected agricultural lands. *Canadian Journal of Plant Science*, *91*(3), 509–517. https://doi.org/10.4141/CJPS10135/ASSET/IMAGES/LARGE/CJPS10135F4.jpeg.

Hossain, A., Krupnik, T. J., Timsina, J., Mahboob, M. G., Chaki, A. K., Farooq, M., Bhatt, R., Fahad, S., & Hasanuzzaman, M. (2020). Agricultural Land Degradation: Processes and Problems Undermining Future Food Security. *Environment, Climate, Plant and Vegetation Growth*, 17–61. https://doi.org/10.1007/978-3-030-49732-3_2.

Jeżowski, S., Mos, M., Buckby, S., Cerazy-Waliszewska, J., Owczarzak, W., Mocek, A., Kaczmarek, Z., & McCalmont, J. P. (2017). Establishment, growth, and yield potential of the perennial grass Miscanthus × Giganteus on degraded coal mine soils. *Frontiers in Plant Science,* 8, 726. https://doi.org/10.3389/FPLS.2017.00726/BIBTEX.

Kairis, O., Karavitis, C., Salvati, L., Kounalaki, A., & Kosmas, K. (2015). *Exploring the Impact of Overgrazing on Soil Erosion and Land Degradation in a Dry Mediterranean Agro-Forest Landscape (Crete, Greece), 29*(3), 360–374. https://doi.org/10.1080/15324982.2014.968691.

Karp, A., & Shield, I. (2008). Bioenergy from plants and the sustainable yield challenge. *New Phytologist*, *179*(1), 15–32. https://doi.org/10.1111/J.1469-8137.2008.02432.X.

Knops, J. M. H., Bradley, K. L., & Wedin, D. A. (2002). Mechanisms of plant species impacts on ecosystem nitrogen cycling. *Ecology Letters*, *5*(3), 454–466. https://doi.org/10.1046/J.1461-0248.2002.00332.X.

Koh, M. Y., & Tinia, T. I. (2011). A review of biodiesel production from Jatropha curcas L. oil. *Renewable and Sustainable Energy Reviews*, *15*(5), 2240–2251. https://doi.org/10.1016/J.RSER.2011.02.013.

Kowalczyk-juśko, A., Mazur, A., Pochwatka, P., Janczak, D., & Dach, J. (2022). Evaluation of the Effects of Using the Giant Miscanthus (Miscanthus & times; Giganteus) Biomass in Various Energy Conversion Processes. *Energies 2022, Vol. 15, Page 3486*, *15*(10), 3486. https://doi.org/10.3390/EN15103486.

Kurganova, I. N., Kuzyakov, Y., Matus, F., Merino, A., Xiang, W., Widyati, E., Sitti Nuroniah, H., Lestari Tata, H., Mindawati, N., Lisnawati, Y., Abdulah, L., Endra Lelana, N., Octavia, D., Prameswari, D., Hendalastuti Rachmat, H., Darwiati, W., Wardani, M., Kalima, T., & van Noordwijk, M. (2022). Soil Degradation Due to Conversion from Natural to Plantation Forests in Indonesia. *Forests 2022, Vol. 13, Page 1913*, *13*(11), 1913. https://doi.org/10.3390/F13111913.

Leontopoulos, S. V., & Arabatzis, G. (2021). The contribution of energy crops to biomass production. *Low Carbon Energy Technologies in Sustainable Energy Systems*, 47–113. https://doi.org/10.1016/B978-0-12-822897-5.00003-1.

Li, C., Wang, Q. H., Xiao, B., & Li, Y. F. (2011). Phytoremediation potential of switchgrass (Panicum virgatum L.) for Cr-polluted soil. *ISWREP 2011 - Proceedings of 2011 International Symposium on Water Resource and Environmental Protection*, *3*, 1731–1734. https://doi.org/10.1109/ISWREP.2011.5893582.

Li, T., Hu, Y., Du, X., Tang, H., Shen, C., & Wu, J. (2014). Salicylic Acid Alleviates the Adverse Effects of Salt Stress in Torreya grandis cv. Merrillii Seedlings by Activating Photosynthesis and Enhancing Antioxidant Systems. *PLOS ONE*, *9*(10), e109492. https://doi.org/10.1371/JOURNAL.PONE.0109492.

Ma, H., Li, X., Wei, M., Zeng, G., Hou, S., Li, D., & Xu, H. (2020). Elucidation of the mechanisms into effects of organic acids on soil fertility, cadmium speciation and ecotoxicity in contaminated soil. *Chemosphere*, *239*, 124706. https://doi.org/10.1016/J.CHEMOSPHERE.2019.124706.

Memon, A. R., Aktoprakligil, D., Özdemir, A., & Vertii, A. (2001). Heavy Metal Accumulation and Detoxification Mechanisms in Plants. *Turkish Journal of Botany*, *25*(3), 111–121.

Mills, A. J., & Fey, M. V. (2004). *Declining soil quality in South Africa: effects of land use on soil organic matter and surface crusting*. https://doi.org/10.10520/EJC97694.

Mohsin, M., Abbas, Q., Zhang, J., Ikram, M., & Iqbal, N. (2019). Integrated effect of energy consumption, economic development, and population growth on CO_2 based environmental degradation: a case of transport sector. *Environmental Science and Pollution Research*, *26*(32), 32824–32835. https://doi.org/10.1007/S11356-019-06372-8/TABLES/7.

Morya, R., Raj, T., Lee, Y., Kumar Pandey, A., Kumar, D., Rani Singhania, R., Singh, S., Prakash Verma, J., & Kim, S. H. (2022). Recent updates in biohydrogen production strategies and life–cycle assessment for sustainable future. *Bioresource Technology,* 366, 128159. https://doi.org/10.1016/J.BIORTECH.2022.128159.

Mu, D., & Ding, C. (2022). Developing a salinity tolerance indicator for tree varieties at challenging sites and urban forests based on inferences of physiological responses: an example of Ulmus pumila. *Trees - Structure and Function*, *36*(2), 593–607. https://doi.org/10.1007/S00468-021-02232-6/FIGURES/3.

Nedjimi, B. (2021). Phytoremediation: a sustainable environmental technology for heavy metals decontamination. *SN Applied Sciences 2021 3:3*, *3*(3), 1–19. https://doi.org/10.1007/S42452-021-04301-4.

Oliver, R. J., Finch, J. W., & Taylor, G. (2009). Second generation bioenergy crops and climate change: a review of the effects of elevated atmospheric CO_2 and drought on water use and the implications for yield. *GCB Bioenergy*, *1*(2), 97–114. https://doi.org/10.1111/J.1757-1707.2009.01011.X.

Pidlisnyuk, V., Shapoval, P., Zgorelec, Ž., Stefanovska, T., & Zhukov, O. (2020). Multiyear phytoremediation and dynamic of foliar metal(loid)s concentration during application of Miscanthus × giganteus Greef et Deu to polluted soil from Bakar, Croatia. *Environmental Science and Pollution Research*, 27(25), 31446–31457. https://doi.org/10.1007/S11356-020-09344-5/FIGURES/6.

Reed, R. L., Sanderson, M. A., Allen, V. G., & Zartman, R. E. (2006). *Cadmium application and pH effects on growth and cadmium accumulation in switchgrass.* 33(7–8), 1187–1203. https://doi.org/10.1081/CSS-120003881.

Reed, S. T., Ayala-Silva, T., Dunn, C. B., Gordon, G. G., & Meerow, A. (2013). Nutrient Uptake of Ornamental Plants Exposed to Arsenic in Hydroponic Solution. *Journal of Agricultural Science*, *5*(12). https://doi.org/10.5539/JAS.V5N12P1.

Schuler, J., Pelkki, M., & Stuhlinger, C. (2009). Biological and Economic Considerations in Establishing a Short-Rotation Bioenergy Plantation. *Journal of the Arkansas Academy of Science*, *63*(1). https://scholarworks.uark.edu/jaas/vol63/iss1/19.

Shrestha, P., Bellitürk, K., & Görres, J. H. (2019). Phytoremediation of Heavy Metal-Contaminated Soil by Switchgrass: A Comparative Study Utilizing Different

Composts and Coir Fiber on Pollution Remediation, Plant Productivity, and Nutrient Leaching. *International Journal of Environmental Research and Public Health* 2019, Vol. 16, Page 1261, 16(7), 1261. https://doi.org/10.3390/IJERPH16071261.

Singh, S., & Verma, J. P. (2019). Bioethanol Production from Different Lignocellulosic Biomass. *Sustainable Biofuel and Biomass: Advances and Impacts,* 281.

Singh, S., Gaurav, A. K., & Verma, J. P. (2020a). Genetically Modified Microbes for Second-Generation Bioethanol Production. In *Fungal Biotechnology and Bioengineering* (pp. 187–198). Springer.

Singh, S., Jaiswal, D. K., Krishna, R., Mukherjee, A., & Verma, J. P. (2020). Restoration of degraded lands through bioenergy plantations. *Restoration Ecology, 28*(2), 263–266.

Singh, S., Jaiswal, D. K., Krishna, R., Mukherjee, A., & Verma, J. P. (2020b). Restoration of degraded lands through bioenergy plantations. *Restoration Ecology,* 28(2), 263–266.

Singh, S., Jaiswal, D. K., Sivakumar, N., & Verma, J. P. (2019). Developing efficient thermophilic cellulose degrading consortium for glucose production from different agro-residues. *Frontiers in Energy Research,* 7, 61.

Singh, S., Kumar, A., Sivakumar, N., & Verma, J. P. (2022). Deconstruction of lignocellulosic biomass for bioethanol production: Recent advances and future prospects. *Fuel,* 327, 125109. https://doi.org/10.1016/J.FUEL.2022.125109.

Singh, S., Verma, J. P., de Araujo Pereira, A. P., & Sivakumar, N. (2018). Production of biofuels and chemicals from lignin | Production of biofuels and chemicals from lignin, Zhen Fang, Richard L. Smith Jr. (Eds.), Springer (2016), p. 442, 10.1007/978-981-10-1965-4, ISBN: 978-981-10-1964-7. *Journal of Cleaner Production,* 203, 966–967. https://doi.org/10.1016/J.JCLEPRO.2018.08.160.

Su, H., Fang, Z., Tsang, P. E., Zheng, L., Cheng, W., Fang, J., & Zhao, D. (2016). Remediation of hexavalent chromium contaminated soil by biochar-supported zero-valent iron nanoparticles. *Journal of Hazardous Materials, 318*, 533–540. https://doi.org/10.1016/J.JHAZMAT.2016.07.039.

Thakur, N. (2017). Organic farming, food quality, and human health: A trisection of sustainability and a move from pesticides to eco-friendly biofertilizers. *Probiotics in Agroecosystem*, 491–515. https://doi.org/10.1007/978-981-10-4059-7_26/COVER.

Tripathi, S., Srivastava, P., Devi, R. S., & Bhadouria, R. (2020). Influence of synthetic fertilizers and pesticides on soil health and soil microbiology. *Agrochemicals Detection, Treatment and Remediation: Pesticides and Chemical Fertilizers*, 25–54. https://doi.org/10.1016/B978-0-08-103017-2.00002-7.

Wong, M. H. (2003). Ecological restoration of mine degraded soils, with emphasis on metal contaminated soils. *Chemosphere, 50*(6), 775–780. https://doi.org/10.1016/S0045-6535(02)00232-1.

Yang, X., Feng, Y., He, Z., & Stoffella, P. J. (2005). Molecular mechanisms of heavy metal hyperaccumulation and phytoremediation. *Journal of Trace Elements in Medicine and Biology, 18*(4), 339–353. https://doi.org/10.1016/J.JTEMB.2005.02.007.

Yanitch, A., Brereton, N. J. B., Gonzalez, E., Labrecque, M., Joly, S., & Pitre, F. E. (2017). Transcriptomic response of purple willow (Salix purpurea) to arsenic stress.

Frontiers in Plant Science, *8*, 1115. https://doi.org/10.3389/FPLS.2017.01115/BIBTEX.

Ye, B., Saito, A., & Minamisawa, K. (2010). *Effect of Inoculation with Anaerobic Nitrogen-Fixing Consortium on Salt Tolerance of Miscanthus sinensis*. 51(2), 243–249. https://doi.org/10.1111/J.1747-0765.2005.TB00028.X.

Yu, X. Z., Gu, J. D., & Xing, L. Q. (2008). Differences in uptake and translocation of hexavalent and trivalent chromium by two species of willows. *Ecotoxicology (London, England)*, *17*(8), 747–755. https://doi.org/10.1007/S10646-008-0224-Y.

Zingore, S., Delve, R. J., Nyamangara, J., & Giller, K. E. (2008). Multiple benefits of manure: The key to maintenance of soil fertility and restoration of depleted sandy soils on African smallholder farms. *Nutrient Cycling in Agroecosystems*, *80*(3), 267–282. https://doi.org/10.1007/S10705-007-9142-2/FIGURES/4.

Zuazo, V. H. D., & Pleguezuelo, C. R. R. (2009). Soil-erosion and runoff prevention by plant covers: A review. *Sustainable Agriculture*, 785–811. https://doi.org/10.1007/978-90-481-2666-8_48/TABLES/4.

Chapter 14

The Role of Soil Microbes in the Reclamation and Restoration of Soil

Sneha Dwivedi
and Alok Kumar Singh*

Laboratory of Microbiology and Plant Pathology, Department of Botany C.M.P. Degree College, University of Allahabad, Prayagraj, India

Abstract

Restoration ecology is the study of the restoration of degraded, damaged, or destroyed ecosystems through active human intervention. Restoration ecology refers to scientific studies that have been developed recently. It is assumed that environmental degradation and population decline are reversible processes. Ecological restoration can improve air quality, reverse deforestation and desertification, slow down biodiversity loss, improve urban environments, and possibly improve human relationships with nature. Recent conceptual advances highlight the interplay between flora, fauna, and soil in restoration ecology. The study of how active human involvement can help restore damaged, destroyed, or degraded ecosystems is known as restoration ecology. Recent scientific research is referred to as "restoration ecology." Environmental deterioration and population decline are thought to be reversible processes. Environmental restoration has the potential to improve urban environments, air quality, biodiversity loss, reversing deforestation and desertification, and human-nature interactions. Current conceptual developments emphasise the interaction of soil, wildlife, and plants in restoration ecology. The time needed for recovery is considerably longer than that for disassembly. Of course restoration is possible, but it takes a very long time. The ecology for recovery starts and quickens the recovery process. Using living things (prokaryotes, fungus, plants, etc.) to clean up a site is known as bioremediation.

* Corresponding Author's Email: dralokkingh@gmail.com.

Keywords: microbial restoration, bioremediation, bacteria, fungi, microbial biomass, soil pollution

1. Introduction

Biotechnology has already played a key role in many fields, advancing our understanding of biological science and its possible applications in different fields (Alisi et al., 2009). Microorganisms are usually related to the biodeterioration of artistic goods, but they can represent a useful tool for the development of new solutions for restoration. A fresh method for conservators to assist in the preservation, protection, and restoration of artworks is the controlled use of microbes as a restorative agent. Existing conservation methods may be ineffective for certain problems or harmful to end users or the environment, thus biotechnology techniques are meant to supplement them rather than replace them (Alisi et al., 2009). Microorganisms in particular have the capacity to decompose, neutralise, and accumulate potentially dangerous organic and inorganic substances. Moreover, it has been claimed that higher plants can filter out these pollutants because of their propensity to accumulate in tissues. Organic pollutants that enter our environment can be broken down (totally or partially), neutralised, and accumulated by natural microorganisms (soil, water, air). a number. The objective of this chapter is to examine various microbial activities that can be harnessed to help prevent pollution and clean up polluted environments. Such microbial activities occur naturally in such environments. It is possible to enhance such activities of microbes through amendments in their habitats (making them more favorable for microbial growth) or by the use of tailored microbes by appropriate genetic engineering technologies. It is now established that although microbes may not remove pollutants entirely, they certainly have the capabilities to reduce pollution significantly.

2. Removal of Toxic Chemicals

Styrene is a fragrant liquid unsaturated hydrocarbon that is primarily used to produce synthetic rubber, resins, and plastics as well as to enhance dry oils. Some fungi, like Candida tropicalis, are capable of assimilation (Nath et al.,

2014). Such styrene-assimilating fungus are grown on a ceramic platform in a laboratory experiment. In comparison to ordinary compost biofilters, the mycelium of the fungi gives the biofilters a huge surface area and a larger capacity to remove contaminants.

3. Removal of Metals

Metal leaching or solubilization by microbes like *Thiobacillus ferrooxidans* is known to remove metals from soils (Oses et al., 2008) Furthermore, microbes can accelerate leaching for recovery using surfactants. Certain surfactants, like the *rhamnolipid* produced by *Pseudomonas aeruginosa*, have cadmium and lead selectivity. *Acinetobacter calcoaceticus* produces emulsan, which can aid in the removal of metal. Many hazardous heavy metals are eliminated from industrial effluents by *Pseudomonas putida*, *Arthrobacter* viscous, and *Citrobacter* species. *Rhizopus arrhizus* removes radioactive elements like uranium and thorium, whereas *Penicillium chrysogenum* can store radium. *Uranium* is accumulated by the yeast Saccharomyces cerevisiae from a diluted solution. The ability of certain fungi from the genera Trichoderma, Aspergillus, *Aureobasidium*, *Ophiostoma*, and *Rhodotorula* to absorb heavy metals has been demonstrated.

4. Removal of Organic Matter

There are many fungi that break down cellulose into simple sugars, including *Aspergillus, Penicillium, Chaetomium, Trichoderma, Fusarium, Stachybotrys*, *Memnoiella, Humicola, Phoma, Cladosporium, Alternaria, Acremonium, Myrothecium,* and *Thielavia.* Members of the basidiomycete families Aphyllophorales and Agaricales mostly break down the lignin found in wood components (Maheswari et al., 2014).

5. Removal of Salinity Stress

The capacity of four endogenous fungi (GM-1, GM-2, GM-3, and GM-4) to promote soybean plants' growth under salt-stress circumstances was examined (Rodriguez et al., 2009). In comparison to the control group, endogenous culture-treated seeds had better rates of seed germination and plant growth. The examination of gibberellin culture filtrates provided more evidence of the fungus' favourable impact on plant growth.

6. Removal of Plant Diseases and Growth of Plants

The use of rhizospheric fungi is an effective and environmentally friendly way to increase plant growth and control various plant diseases. The three predominant fungi (PNF1, PNF2, and PNF3) isolated from the rhizosphere soil of peanut plants were tested for their effect on stimulating the growth of sesame seeds (Karmakar, et al., 2013). Among these isolates, PNF2 significantly increased shoot length and fresh seedling weight compared to controls. Analysis of the fungal culture filtrate showed a higher concentration of indole acetic acid in PNF2 compared to other isolates.

Arbuscular mycorrhizal fungi (AMF) are helpful fungi that coexist in symbiosis with several land plants. By enhancing the soil's characteristics, arbuscular mycorrhizal fungi can encourage plant growth both under typical and stressful circumstances. Through modifying plant shape, nutrition, and physiological levels, arbuscular mycorrhizal fungal colonies stimulate plant development and boost resilience to diverse abiotic challenges. *Ocimum basilicum* is protected against salt by arbuscular mycorrhiza inoculation, which improves mineral uptake, chlorophyll production, and water use efficiency. It has been demonstrated that tomato plants inoculated with branched mycorrhizal fungi develop more quickly than control plants because their leaves have more surface area, nitrogen, potassium, calcium, and phosphorus.

7. Biocontrol

Certain fungus, including *Cordyceps melothac, Metarhizium anisopliae,* and *Empusa sepulchrasis,* can be utilised to manage some insect infestations (Pimentel, 2011). The majority of the others are spore-forming parasites of specific insects. Utilizative mould, the citrus rust mite, and the Colorado potato beetle can all be used to manage carnivorous insects. In the *mycelium,* this kind of fungus creates rings that capture and suffocate nematodes as they attempt to pass through. Then they start eating nematodes.

8. Myconematicides

Three forms of neuropathogenic fungi nematodes, trapping fungi (*Arthrobotrys, Dactylella*), endoparasites (*Hirsutella, Meria*), and very particular egg parasites are based on the nature of fungal biocontrol agents (*Datylella*). Royal 300 R (*Arthrobotrys robata*) and Royal 350 R (*Arthrobotrys superb*) are the most popular and widely used myconematicides (Vega et al., 2008).

Conclusion

Microbiological restoration, in particular through the use of native fungi, is an important resource for community restoration projects that aim to restore areas severely affected by human-induced hazards that have resulted in ecosystem destruction due to erosion, decay, natural disasters, pollution, or mismanagement. Fungi can colonize and regenerate soils, vegetation, forest floors, and completely destroyed habitats. The root system of a fungus is called mycelium. It can grow in soil and moist environments such as rotting tree trunks to find and connect nutrients. Mycelium is one of the longest-living organisms in nature. It is waterproof and fireproof, making it an ideal tool for environmental recovery and disaster response. Plants are based on micro cleaning, which recognizes the natural ability of fungi as decomposers and decomposers to regenerate ecosystems. In their laboratories, scientists can hone particular fungi to break down waste products like polypropylene face masks, plastic gloves, and sheet metal. Businesses like Mycocycle are also developing bioreactors that enable fungi to break down materials in a controlled environment. In order to increase the ability of plants to recognise and adapt to environmental stresses like drought, salinity, heat, cold, and heavy metals, fungi play a fundamental role in various physiological processes, mineral and water uptake, chemical change, and the biosynthesis of compounds like auxins, lignan, and ethylene. The majority of experimental breakdown mechanisms take place before mould or fruiting bodies appear. During a few weeks, the fungus normally entirely digests or absorbs the poisons or waste products.

References

Alisi, C., Musella, R., Tasso, F., Ubaldi, C., Manzo, S., Cremisini, C., & Sprocati, A. R. (2009). Bioremediation of diesel oil in a co-contaminated soil by bioaugmentation with a microbial formula tailored with native strains selected for heavy metals resistance. *Science of the Total Environment*, 407(8), https://doi.org/10.1016/j.scitotenv.2009.01.011.

Karmakar, R., Sunil, K., & Prakash, H. S. (2013). Antioxidant, anti-inflammatory, antimicrobial and cytotoxic properties of fungal endophytes from Garcinia species. *International Journal of Pharmacy and Pharmaceutical Sciences,* 5(3), 889–897.

Maheshwari, V. L., Patil, M. P., Patil, R. H., & Patil, S. G. (2014). Endophytic Mycoflora of Indian medicinal plant, Terminalia Arjuna and their biological activities. International *Journal of Biotechnology for Wellness Industries,* 3(2), 53–61. https://doi.org/10.6000/1927-3037.2014.03.02.3.

Nath, A., Pathak, J., & Joshi, S. R. (2014). Bioactivity assessment of endophytic fungi associated with Centella asiatica and Murraya Koengii. *Journal of Applied Biology and Biotechnology*, 2(5), 0–1.

Oses, R., Valenzuela, S., Freer, J., Sanfuentes, E., & Rodriguez, J. (2008). Fungal endophytes in xylem of healthy Chilean trees and their possible role in early wood decay. *Fungal Diversity*, 33(7), 77–86.

Pimentel, D. (2011). *Introduction: Nonnative species in the world. In Biological Invasions.* CRC Press, 15–24.

Ranalli, G., Alfano, G., Belli, C., Lustrato, G., Colombini, M. P., Bonaduce, I., Zanardini, E., Abbruscato, P., Cappitelli, F., & Sorlini, C. (2005). Biotechnology applied to cultural heritage: Biorestoration of frescoes using viable bacterial cells and enzymes. *Journal of Applied Microbiology*, 98(1), 73–83. https://doi.org/10.1111/j.1365-2672.2004.02429.x.

Ranalli, G., Chiavarini, M., Guidetti, V., Marsala, F., Matteini, M., Zanardini, E., & Sorlini, C. (1997). The use of microorganisms for the removal of sulphates on artistic stoneworks. *International Biodeterioration and Biodegradation,* 40(2–4), 255–261. https://doi.org/10.1016/S0964-8305(97)00054-1.

Rodriguez, R. J., White, J. F., Jr., Arnold, A. E., & Redman, R. S. (2009) Fungal endophytes: Diversity and functional roles. *New Phytologist*, 182(2), 314–330. https://doi.org/10.1111/j.1469-8137.2009.02773.x.

Sorlini, C., Gulotta, D., Ranalli, G., Zanardini, E., Cappitelli, F., Toniolo, L., & Sansonetti, A. (2007). *Advantages of using microbial technology,* 73(17) (pp. 5671–5675).

Vega, F. E., Posada, F., Catherine Aime, M. C., Pava-Ripoll, M., Infante, F., & Rehner, S. A. (2008). Entomopathogenic fungal endophytes. *Biological Control,* 46(1), 72–82. https://doi.org/10.1111/j.1469-8137.2009.02773.x.

About the Editors

Dr. Rajani Srivastava, MSc, PhD (Botany, Sp. Ecology), is presently affiliated as an Assistant Professor at Institute of Environment & Sustainable Development, RGSC, BHU, India. Research interest includes Restoration Ecology, agroecosystem ecology, weed ecology and sustainable agriculture. Have published 3 National and 2 International books and more than 50 scholarly publications in international journals and attended >65 National & International seminars/conferences/workshops.

Dr. Pankaj Kumar, working as professor in Botany. Having a teaching and research experience of over twenty years. Specialized in Ecology and Environment. He is having more than thirty five publications in various national and international journals.

Index

G

H

I

L

M

N